Table of ? Contents

D1120449

**Thinking and the
Mathematics Standards....... 3
Introduction 5**

1 **Problem Solving.............. 29**

2 **Number Patterns
   and Relationships 37**

3 **Whole Number and
   Decimal Place Value 45**

4 **Decimal Operations......... 53**

5 **Fraction Concepts 61**

6 **Fraction Addition
   and Subtraction 69**

7 **Fraction Multiplication
   and Division.................... 77**

8 **Percents 85**

9 **Relating Fractions,
   Decimals, and Percents ... 93**

10 **Ratios, Rates, and
    Proportions 101**

11 **Statistics,
    Data Analysis,
    and Graphing 109**

12 **Algebra: Integers and
    Integer Operations........ 117**

13 **Algebra: Solving Equations
    and Inequalities 125**

14 **Geometry Concepts 133**

15 **Geometry
    and Measurement......... 141**

16 **Probability 149**

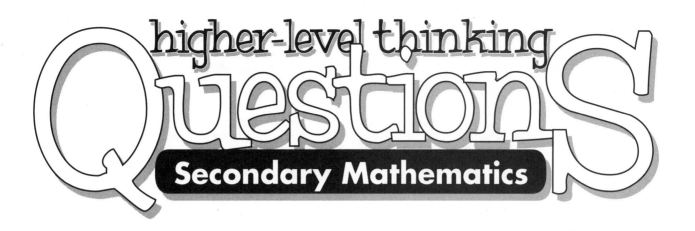

higher-level thinking Questions
Secondary Mathematics

questions by
Robyn Silbey

created and designed by
Miguel Kagan

layout by
Miles Richey

illustrated by
**Erin Kant and
Celso Rodriguez**

Kagan

Kagan Publishing
P.O. Box 72008
San Clemente, CA 92673-2008
1(800) 933-2667
www.KaganOnline.com

ISBN 978-1-879097-86-5

Thinking and the Mathematics Standards

How does this book support the National Mathematics standards? What do the mathematics standards call for? The National Council of Teachers of Mathematics released its first standards document in 1989. The document was replaced by a 1991 document. In April 2000, NCTM released the Principles and Standards for School Mathematics. In all three documents, the message is consistent: Mathematics curricula and instruction are to move beyond memorization, rote learning, and application of predetermined procedures. The standards call on teachers to work toward a deeper conceptual understanding and foster mathematical reasoning.

The mathematics standards include an array of higher-level thinking skills. We are to prepare students for the rapidly changing high technology of the 21st century job world, which will require many different creative uses of mathematics. The standards encourage us to teach students to create their own new, alternative algorithms. They expect us to challenge students to be creative mathematicians, generating new facts and new ways to organize, analyze, synthesize, and interpret existing facts. We cannot be sure of what kinds of problems our students will solve or what kinds of data they will need to interpret, but we can be certain that mathematical reasoning, communication with and about math, and a mathematical problem-solving orientation will serve them well.

We must teach mathematics for understanding. Students must
- discover a variety of ways to compute
- communicate about math
- use math to communicate
- demonstrate a sense of number
- develop the ability to solve problems creatively
- expand mathematical ways of thinking

To be successful, students need to think mathematically—to approach problems in many ways. This shift can be accomplished if we make a corresponding shift in instruction, teaching students how to develop higher-order thinking skills. We need to move instruction away from direct instruction and individual worksheet practice toward active engagement through cooperative thinking, discussions, and mental investigations. Communication about and with mathematical concepts is at the core of the new vision.

If students are to be engaged in the exploration of mathematical relations, to come to know and understand mathematics, the best set of strategies involves cooperative interaction. And if we are to have our students interacting as they explore their mathematics, the Kagan Structures offer the advantage of having students all engaged and all accountable.

The Kagan Cooperative Learning Structures are empowering for both teachers and students. For teachers, the structures are easy-to-learn instructional strategies. For students, the structures provide alternative windows into the curriculum. In addition, students are learning to coordinate efforts with others, all others, creating an inclusive classroom. In the process, students acquire mutual respect, shared leadership, and a sense of "our math class."

> "I had six
> honest serving men
> They taught me all I knew:
> Their names were Where
> and What and When
> and Why and How and
> Who.

— Rudyard Kipling

Higher-Level Thinking Questions for Secondary Mathematics
Kagan Publishing • 1 (800) 933-2667 • www.KaganOnline.com

Introduction

In your hands you hold a powerful book. It is a member of a series of transformative blackline activity books. Between the covers, you will find questions, questions, and more questions! But these are no ordinary questions. These are the important kind—higher-level thinking questions—the kind that stretch your students' minds; the kind that release your students' natural curiosity about the world; the kind that rack your students' brains; the kind that instill in your students a sense of wonderment about your curriculum.

But we are getting a bit ahead of ourselves. Let's start from the beginning. Since this is a book of questions, it seems only appropriate for this introduction to pose a few questions—about the book and its underlying educational philosophy. So Mr. Kipling's Six Honest Serving Men, if you will, please lead the way:

What?
What are higher-level thinking questions?

This is a loaded question (as should be all good questions). Using our analytic thinking skills, let's break this question down into two smaller questions: 1) What is higher-level thinking? and 2) What are questions? When we understand the types of thinking skills and the types of questions, we can combine the best of both worlds, crafting beautiful questions to generate the range of higher-level thinking in our students!

Types of Thinking

There are many different types of thinking. Some types of thinking include:

- applying
- associating
- comparing
- contrasting
- defining
- elaborating
- empathizing
- experimenting
- generalizing
- investigating
- making analogies
- planning
- prioritizing
- recalling
- reflecting
- reversing
- sequencing
- summarizing
- synthesizing
- assessing
- augmenting
- connecting
- decision-making
- drawing conclusions
- eliminating
- evaluating
- explaining
- inferring consequences
- inventing
- memorizing
- predicting
- problem-solving
- reducing
- relating
- role-taking
- substituting
- symbolizing
- understanding
- thinking about thinking (metacognition)

This is quite a formidable list. It's nowhere near complete. Thinking is a big, multifaceted phenomenon. Perhaps the most widely recognized system for classifying thinking and classroom questions is Benjamin Bloom's Taxonomy of Thinking Skills. Bloom's Taxonomy classifies thinking skills into six hierarchical levels. It begins with the lower levels of thinking skills and moves up to higher-level thinking skills: 1) Knowledge, 2) Comprehension, 3) Application, 4) Analysis, 5) Synthesis, 6) Evaluation. See Bloom's Taxonomy on the following page.

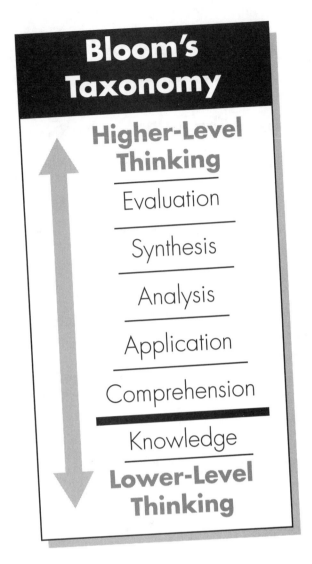

Bloom's Taxonomy

Higher-Level Thinking

Evaluation

Synthesis

Analysis

Application

Comprehension

Knowledge

Lower-Level Thinking

In education, the term "higher-level thinking" often refers to the higher levels of Mr. Bloom's taxonomy. But Bloom's Taxonomy is but one way of organizing and conceptualizing the various types of thinking skills.

There are many ways we can cut the thinking skills pie. We can alternatively view the many different types of thinking skills as, well…many different skills. Some thinking skills may be hierarchical. Some may be interrelated. And some may be relatively independent.

In this book, we take a pragmatic, functional approach. Each type of thinking skill serves a different function. So called "lower-level" thinking skills are very useful for certain purposes. Memorizing and understanding information

are invaluable skills that our students will use throughout their lives. But so too are many of the "higher-level" thinking skills on our list. The more facets of students' thinking skills we develop, the better we prepare them for lifelong success.

Because so much classroom learning heretofore has focused on the "lower rungs" of the thinking skills ladder—knowledge and comprehension, or memorization and understanding—in this series of books we have chosen to focus on questions to generate "higher-level" thinking. This book is an attempt to correct the imbalance in the types of thinking skills developed by classroom questions.

Types of Questions

As we ask questions of our students, we further promote cognitive development when we use Fat questions, Low-Consensus questions, and True questions.

Fat Questions vs. Skinny Questions

Skinny questions are questions that require a skinny answer. For example, after reading a poem, we can ask: "Did you like the poem?" Even though this question could be categorized as an Evaluation question—Bloom's highest level of thinking— it can be answered with one monosyllabic word: "Yes" or "No." How much thinking are we actually generating in our students?

We can reframe this question to make it a fat question: "What things did you like about the poem? What things did you dislike?" Notice no short answer will do. Answering this fattened-up question requires more elaboration. These fat questions presuppose not that there is only one thing but things plural that the student liked and things that she did not like. Making things plural is one way to make skinny questions fat. Students stretch their minds to come up with multiple ideas or solutions. Other easy ways to

Higher-Level Thinking Questions for Secondary Mathematics
Kagan Publishing • 1 (800) 933-2667 • www.KaganOnline.com

make questions fat is to add "Why or why not?" or "Explain" or "Describe" or "Defend your position" to the end of a question. These additions promote elaboration beyond a skinny answer. Because language and thought are intimately intertwined, questions that require elaborate responses stretch students' thinking: They grapple to articulate their thoughts.

The type of questions we ask impact not just the type of thinking we develop in our students, but also the depth of thought. Fat questions elicit fat responses. Fat responses develop both depth of thinking and range of thinking skills. The questions in this book are designed to elicit fat responses—deep and varied thinking.

High-Consensus Questions vs. Low-Consensus Questions

A high-consensus question is one to which most people would give the same response, usually a right or wrong answer. After learning about sound, we can ask our students: "What is the name of a room specially designed to improve acoustics for the audience?" This is a high-consensus question. The answer (auditorium) is either correct or incorrect.

Compare the previous question with a low-consensus question: "If you were going to build an auditorium, what special design features would you take into consideration?" Notice, to the low-consensus question there is no right or wrong answer. Each person formulates his or her unique response. To answer, students must apply what they learned, use their ingenuity and creativity.

High-consensus questions promote convergent thinking. With high-consensus questions we strive to direct students *what to think*. Low-consensus questions promote divergent thinking, both critical and creative. With low-consensus

questions we strive to develop students' *ability to think*. The questions in this book are low-consensus questions designed to promote independent, critical and creative thought.

True Questions vs. Review Questions

We all know what review questions are. They're the ones in the back of every chapter and unit. Review questions ask students to regurgitate previously stated or learned information. For example, after learning about the rain forest we may ask: "What percent of the world's oxygen does the rain forest produce?" Students can go back a few pages in their books or into their memory banks and pull out the answer. This is great if we are working on memorization skills, but does little to develop "higher-order" thinking skills.

True questions, on the other hand, are meaningful questions—questions to which we do not know the answer. For example: "What might happen if all the world's rain forests were cut down?" This is a hypothetical; we don't know the answer but considering the question forces us to think. We infer some logical consequences based on what we know. The goal of true questions is not a correct answer, but the thinking journey students take to create a meaningful response. True questions are more representative of real life. Seldom is there a black and white answer. In life, we struggle with ambiguity, confounding variables, and uncertain outcomes. There are millions of shades of gray. True questions prepare students to deal with life's uncertainties.

When we ask a review question, we know the answer and are checking to see if the student does also. When we ask a true question, it is truly a question. We don't necessarily know the answer and neither does the student. True questions are

Education is not the filling of a pail, but the lighting of a fire.

— William Butler Yeats

Types of Questions

Skinny ➡	**Fat**
• Short Answer	• Elaborated Answer
• Shallow Thinking	• Deep Thinking

High-Consensus ➡	**Low-Consensus**
• Right or Wrong Answer	• No Single Correct Answer
• Develops Convergent Thinking	• Develops Divergent Thinking
• "What" to Think	• "How" to Think

Review ➡	**True**
• Asker Knows Answer	• Asker Doesn't Know Answer
• Checking for Correctness	• Invitation to Think

often an invitation to think, ponder, speculate, and engage in a questioning process.

We can use true questions in the classroom to make our curriculum more personally meaningful, to promote investigation, and awaken students' sense of awe and wonderment in what we teach. Many questions you will find in this book are true questions designed to make the content provocative, intriguing, and personally relevant.

The box above summarizes the different types of questions. The questions you will find in this book are a move away from skinny, high-consensus, review questions toward fat, low-consensus true questions. As we ask these types of questions in our class, we transform even mundane content into a springboard for higher-level thinking. As we integrate these question gems into our daily lessons, we create powerful learning experiences. *We do not fill our students' pails with knowledge; we kindle their fires to become lifetime thinkers.*

Why?

Why should I use higher-level thinking questions in my classroom?

As we enter the new millennium, major shifts in our economic structure are changing the ways we work and live. The direction is increasingly toward an information-based, high-tech economy. The sum of our technological information is exploding. We could give you a figure how rapidly information is doubling, but by the time you read this, the number would be outdated! No kidding.

But this is no surprise. This is our daily reality. We see it around us everyday and on the news: cloning, gene manipulation, e-mail, the Internet, Mars rovers, electric cars, hybrids, laser surgery, CD-ROMs, DVDs. All around us we see the wheels of progress turning: New discoveries, new technologies, a new societal knowledge and information base. New jobs are being created to-

Higher-Level Thinking Questions for Secondary Mathematics
Kagan Publishing • 1 (800) 933-2667 • www.KaganOnline.com

day in fields that simply didn't exist yesterday.

How do we best prepare our students for this uncertain future—a future in which the only constant will be change? As we are propelled into a world of ever-increasing change, what is the relative value of teaching students facts versus thinking skills? This point becomes even more salient when we realize that students cannot master everything, and many facts will soon become obsolete. Facts become outdated or irrelevant. Thinking skills are for a lifetime. Increasingly, how we define educational success will be away from the quantity of information mastered. Instead, we will define success as our students' ability to generate questions, apply, synthesize, predict, evaluate, compare, categorize.

If we as a professionals are to proactively respond to these societal shifts, thinking skills will become central to our curriculum. Whether we teach thinking skills directly, or we integrate them into our curriculum, the power to think is the greatest gift we can give our students!

We believe the questions you will find in this book are a step in the direction of preparing students for lifelong success. The goal is to develop independent thinkers who are critical and creative, regardless of the content. We hope the books in this series are more than sets of questions. We provide them as a model approach to questioning in the classroom.

On pages 10 and 11, you will find Questions to Engage Students' Thinking Skills. These pages contain numerous types of thinking and questions designed to engage each thinking skill. As you make your own questions for your students with your own content, use these question starters to help you frame

> ## Virtually the only predictable trend is continuing change.
> — Dr. Linda Tsantis, Creating the Future

your questions to stimulate various facets of your students' thinking skills. Also let your students use these question starters to generate their own higher-level thinking questions about the curriculum.

Who?
Who is this book for?

This book is for you and your students, but mostly for your students. It is designed to help make your job easier. Inside you will find hundreds of ready-to-use reproducible questions. Sometimes in the press for time we opt for what is easy over what is best. These books attempt to make easy what is best. In this treasure chest, you will find hours and hours of timesaving ready-made questions and activities.

Place Higher-Level Thinking In Your Students' Hands

As previously mentioned, this book is even more for your students than for you. As teachers, we ask a tremendous number of questions. Primary teachers ask 3.5 to 6.5 questions per minute! Elementary teachers average 348 questions a day. How many questions would you predict our students ask? Researchers asked this question. What they found was shocking: Typical students ask approximately one question per month.* One question per month!

Although this study may not be representative of your classroom, it does suggest that in general, as teachers we are missing out on a very powerful force—student-generated questions. The capacity to answer higher-level thinking questions is

---

* Myra & David Sadker, "Questioning Skills" in *Classroom Teaching Skills*, 2nd ed. Lexington, MA: D.C. Heath & Co., 1982.

Questions to Engage Students' Thinking Skills

Analyzing
- How could you break down…?
- What components…?
  - What qualities/characteristics…?

Applying
- How is _____ an example of…?
- What practical applications…?
- What examples…?
- How could you use…?
- How does this apply to…?
- In your life, how would you apply…?

Assessing
- By what criteria would you assess…?
- What grade would you give…?
- How could you improve…?

Augmenting/Elaborating
- What ideas might you add to…?
- What more can you say about…?

Categorizing/Classifying/Organizing
- How might you classify…?
- If you were going to categorize…?

Comparing/Contrasting
- How would you compare…?
- What similarities…?
- What are the differences between…?
- How is _____ different…?

Connecting/Associating
- What do you already know about…?
- What connections can you make between…?
- What things do you think of when you think of…?

Decision-Making
- How would you decide…?
- If you had to choose between…?

Defining
- How would you define…?
- In your own words, what is…?

Describing/Summarizing
- How could you describe/summarize…?
- If you were a reporter, how would you describe…?

Determining Cause/Effect
- What is the cause of…?
- How does _____ effect _____?
- What impact might…?

Drawing Conclusions/ Inferring Consequences
- What conclusions can you draw from…?
- What would happen if…?
- What would have happened if…?
- If you changed _____, what might happen?

Eliminating
- What part of _____ might you eliminate?
- How could you get rid of…?

Evaluating
- What is your opinion about…?
- Do you prefer…?
- Would you rather…?
- What is your favorite…?
- Do you agree or disagree…?
- What are the positive and negative aspects of…?
- What are the advantages and disadvantages…?
- If you were a judge…?
- On a scale of 1 to 10, how would you rate…?
- What is the most important…?
- Is it better or worse…?

Explaining
- How can you explain…?
- What factors might explain…?

Higher-Level Thinking Questions for Secondary Mathematics
Kagan Publishing • 1 (800) 933-2667 • www.KaganOnline.com

Experimenting
• How could you test…?
• What experiment could you do to…?

Generalizing
• What general rule can…?
• What principle could you apply…?
• What can you say about all…?

Interpreting
• Why is ____ important?
• What is the significance of…?
• What role…?
• What is the moral of…?

Inventing
• What could you invent to…?
• What machine could…?

Investigating
• How could you find out more about…?
• If you wanted to know about…?

Making Analogies
• How is ____ like ____?
• What analogy can you invent for…?

Observing
• What observations did you make about…?
• What changes…?

Patterning
• What patterns can you find…?
• How would you describe the organization of…?

Planning
• What preparations would you…?

Predicting/Hypothesizing
• What would you predict…?
• What is your theory about…?
• If you were going to guess…?

Prioritizing
• What is more important…?
• How might you prioritize…?

Problem-Solving
• How would you approach the problem?
• What are some possible solutions to…?

Reducing/Simplifying
• In a word, how would you describe…?
• How can you simplify…?

Reflecting/Metacognition
• What would you think if…?
• How can you describe what you were thinking when…?

Relating
• How is ____ related to ____?
• What is the relationship between…?
• How does ____ depend on ____?

Reversing/Inversing
• What is the opposite of…?

Role-Taking/Empathizing
• If you were (someone/something else)…?
• How would you feel if…?

Sequencing
• How could you sequence…?
• What steps are involved in…?

Substituting
• What could have been used instead of…?
• What else could you use for…?
• What might you substitute for…?
• What is another way…?

Symbolizing
• How could you draw…?
• What symbol best represents…?

Synthesizing
• How could you combine…?
• What could you put together…?

a wonderful skill we can give our students, as is the skill to solve problems. Arguably more important skills are the ability to find problems to solve and formulate questions to answer. If we look at the great thinkers of the world—the Einsteins, the Edisons, the Freuds—their thinking is marked by a yearning to solve tremendous questions and problems. It is this questioning process that distinguishes those who illuminate and create our world from those who merely accept it.

> **Asking a good question requires students to think harder than giving a good answer.**
> — Robert Fisher, Teaching Children to Learn

Reflect on this analogy: If we wanted to teach our students to catch and throw, we could bring in one tennis ball and take turns throwing it to each student and having them throw it back to us. Alternatively, we could bring in twenty balls and have our students form small groups and have them toss the ball back and forth to each other. Picture the two classrooms: One with twenty balls being caught at any one moment, and the other with just one. In which class would students better and more quickly learn to catch and throw?

Make Learning an Interactive Process

Higher-level thinking is not just something that occurs between students' ears! Students benefit from an interactive process. This basic premise underlies the majority of activities you will find in this book.

As students discuss questions and listen to others, they are confronted with differing perspectives and are pushed to articulate their own thinking well beyond the level they could attain on their own. Students too have an enormous capacity to mediate each other's learning. When we heterogeneously group students to work together, we create an environment to move students through their zone of proximal development. We also provide opportunities for tutoring and leadership. Verbal interaction with peers in cooperative groups adds a dimension to questions not available with whole-class questions and answers.

The same is true with thinking skills. When we make our students more active participants in the learning process, they are given dramatically more opportunities to produce their own thought and to strengthen their own thinking skills. Would you rather have one question being asked and answered at any one moment in your class, or twenty? Small groups mean more questioning and more thinking. Instead of rarely answering a teacher question or rarely generating their own question, asking and answering questions becomes a regular part of your students' day. It is through cooperative interaction that we truly turn our classroom into a higher-level think tank. The associated personal and social benefits are invaluable.

When?
When do I use higher-level thinking questions?

Do I use these questions at the beginning of the lesson, during the lesson, or after? The answer, of course, is all of the above.

Use these questions or your own thinking questions at the beginning of the lesson to provide a motivational set for the lesson. Pique students' interest about the content with some provocative questions: "What would happen if we didn't have gravity?" "Why did Pilgrims get along with some Native Americans, but not others?" "What do you think this book will be about?" Make the content personally relevant by bringing in students' own knowledge, experiences, and feelings about the content: "What do you know about spiders?" "What things do you like about mystery stories?" "How would you feel if explorers invaded your land and killed your family?" "What do you wonder about electricity?"

Use the higher-level thinking questions throughout your lessons. Use the many questions and activities in this book not as a replacement of your curriculum, but as an additional avenue to explore the content and stretch students' thinking skills.

Use the questions after your lesson. Use the higher-level thinking questions, a journal writing activity, or the question starters as an extension activity to your lesson or unit.

Or just use the questions as stand-alone sponge activities for students or teams who have finished their work and need a challenging project to work on.

It doesn't matter when you use them, just use them frequently. As questioning becomes a habitual part of the classroom day, students' fear of asking silly questions is diminished. As the ancient Chinese proverb states, "Those who ask a silly question may seem a fool for five minutes, but those who do not ask remain a fool for life."

> ## The important thing is to never stop questioning.
> — Albert Einstein

As teachers, we should make a conscious effort to ensure that a portion of the many questions we ask on a daily basis are those that move our students beyond rote memorization. When we integrate higher-level thinking questions into our daily lessons, we transform our role from transmitters of knowledge to engineers of learning.

Where?
Where should I keep this book?

Keep it close by. Inside there are 16 sets of questions. Pull it out any time you teach these topics or need a quick, easy, fun activity or journal writing topic.

How?
How do I get the most out of this book?

In this book you will find 16 topics arranged alphabetically. For each topic there are reproducible pages for: 1) 16 Question Cards, 2) a Journal Writing activity page, 3) and a Question Starters activity page.

1. Question Cards

The Question Cards are truly the heart of this book. There are numerous ways the Question Cards can be used. After the other activity pages are introduced, you will find a description of a variety of engaging formats to use the Question Cards.

Specific and General Questions

Some of the questions provided in this book series are content-specific and others are content-free. For example, the literature questions in the Literature books are content-specific. Questions for the Great Kapok Tree deal specifically with that literature selection. Some language arts questions in the Language Arts book, on the other hand, are content-free. They are general questions that can be used over and over again with new content. For example, the Book Review questions can be used after reading any book. The Story Structure questions can be used after reading any story. You can tell by glancing at the title of the set and some of the questions whether the set is content-specific or content-free.

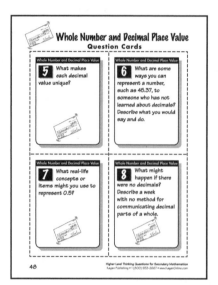

A Little Disclaimer

Not all of the "questions" on the Question Cards are actually questions. Some instruct students to do something. For example, "Compare and contrast…" We can also use these directives to develop the various facets of students' thinking skills.

The Power of Think Time

As you and your students use these questions, don't forget about the power of Think Time! There are two different think times. The first is the time between the question and the response. The second is the time between the response and feedback on the response. Think time has been shown to greatly enhance the quality of student thinking. If students are not pausing for either think time, or doing it too briefly, emphasize its importance. Five little seconds of silent think time after the question and five more seconds before feedback are proven, powerful ways to promote higher-level thinking in your class.

Use Your Question Cards for Years

For attractive Question Cards that will last for years, photocopy them on color card-stock paper and laminate them. To save time, have the Materials Monitor from each team pick up one card set, a pair of scissors for the team, and an envelope or rubber band. Each team cuts out their own set of Question Cards. When they are done with the activity, students can place the Question Cards in the envelope and write the name of the set on the envelope or wrap the cards with a rubber band for storage.

Higher-Level Thinking Questions for Secondary Mathematics
Kagan Publishing • 1 (800) 933-2667 • www.KaganOnline.com

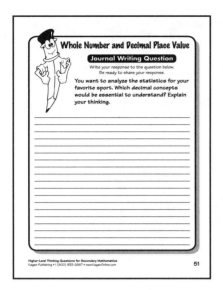

2. Journal Question

The Journal Writing page contains one of the 16 questions as a journal writing prompt. You can substitute any question, or use one of your own. The power of journal writing cannot be overstated. The act of writing takes longer than speaking and thinking. It allows the brain time to make deep connections to the content. Writing requires the writer to present his or her response in a clear, concise language. Writing develops both strong thinking and communication skills.

A helpful activity before journal writing is to have students discuss the question in pairs or in small teams. Students discuss their ideas and what they plan to write. This little prewriting activity ignites ideas for those students who stare blankly at their Journal Writing page. The interpersonal interaction further helps students articulate what they are thinking about the topic and invites students to delve deeper into the topic.

Tell students before they write that they will share their journal entries with a partner or with their team. This motivates many students to improve their entry. Sharing written responses also promotes flexible thinking with open-ended questions, and allows students to hear their peers' responses, ideas and writing styles.

Have students keep a collection of their journal entries in a three-ring binder. This way you can collect them if you wish for assessment or have students go back to reflect on their own learning. If you are using questions across the curriculum, each subject can have its own journal or own section within the binder. Use the provided blackline on the following page for a cover for students' journals or have students design their own.

3. Question Starters

The Question Starters activity page is designed to put the questions in the hands of your students. Use these question starters to scaffold your students' ability to write their own thinking questions. This page includes eight question starters to direct students to generate questions across the levels and types of thinking. This Question Starters activity page can be used in a few different ways:

Individual Questions

Have students independently come up with their own questions. When done, they can trade their questions with a partner. On a separate sheet of paper students answer their partners' questions. After answering, partners can share how they answered each other's questions.

JOURNAL

My Best Thinking

_____

This Journal Belongs to

Higher-Level Thinking Questions for Secondary Mathematics
Kagan Publishing • 1 (800) 933-2667 • www.KaganOnline.com

Pair Questions

Students work in pairs to generate questions to send to another pair. Partners take turns writing each question and also take turns recording each answer. After answering, pairs pair up to share how they answered each other's questions.

Team Questions

Students work in teams to generate questions to send to another team. Teammates take turns writing each question and recording each answer. After answering, teams pair up to share how they answered each other's questions.

Teacher-Led Questions

For young students, lead the whole class in coming up with good higher-level thinking questions.

Teach Your Students About Thinking and Questions

An effective tool to improve students' thinking skills is to teach students about the types of thinking skills and types of questions. Teaching students about the types of thinking skills improves their metacognitive abilities. When students are aware of the types of thinking, they may more effectively plan, monitor, and evaluate their own thinking. When students understand the types of questions and the basics of question construction, they are more likely to create effective higher-level thinking questions. In doing so they develop their own thinking skills and the thinking of classmates as they work to answer each other's questions.

Table of Activities

The Question Cards can be used in a variety of game-like formats to forge students' thinking skills. They can be used for cooperative team and pair work, for whole-class questioning, for independent activities, or at learning centers. On the following pages you will find numerous excellent options to use your Question Cards. As you use the Question Cards in this book, try the different activities listed below to add novelty and variety to the higher-level thinking process.

Team Activities
1. Question Commander 18
2. Fan-N-Pick.......................... 20
3. Spin-N-Think 20
4. Three-Step Interview 21
5. Team Discussion 21
6. Think-Pair-Square 22
7. Question-Write-RoundRobin 22

Class Activities
1. Mix-Pair-Discuss..................... 23
2. Think-Pair-Share 23
3. Inside-Outside Circle 24
4. Question & Answer.................. 24
5. Numbered Heads Together 25

Pair Activities
1. RallyRobin 25
2. Pair Discussion 26
3. Question-Write-Share-Discuss 26

Individual Activities
1. Journal Writing....................... 27
2. Independent Answers 27

Learning Centers
1. Question Card Center 28
2. Journal Writing Center............. 28
3. Question Starters Center 28

Activities

team activity #1

Question Commander

Preferably in teams of four, students shuffle their Question Cards and place them in a stack, questions facing down, so that all teammates can easily reach the Question Cards. Give each team a Question Commander set of instructions (blackline provided on following page) to lead them through each question.

Student One becomes the Question Commander for the first question. The Question Commander reads the question aloud to the team, then asks the teammates to think about the question and how they would answer it. After the think time, the Question Commander selects a teammate to answer the question. The Question Commander can spin a spinner or roll a die to select who will answer. After the teammate gives the answer, Question Commander again calls for think time, this time asking the team to think about the answer. After the think time, the Question Commander leads a team

discussion in which any teammember can contribute his or her thoughts or ideas to the question, or give praise or reactions to the answer.

When the discussion is over, Student Two becomes the Question Commander for the next question.

Question Commander

Higher-Level Thinking Questions for Secondary Mathematics
Kagan Publishing • 1 (800) 933-2667 • www.KaganOnline.com

Question Commander
Instruction Cards

Question Commander

1. Ask the Question: Question Commander reads the question to the team.

2. Think Time: "Think of your best answer."

3. Answer the Question: The Question Commander selects a teammate to answer the question.

4. Think Time: "Think about how you would answer differently or add to the answer."

5. Team Discussion: As a team, discuss other possible answers or reactions to the answer given.

Question Commander

1. Ask the Question: Question Commander reads the question to the team.

2. Think Time: "Think of your best answer."

3. Answer the Question: The Question Commander selects a teammate to answer the question.

4. Think Time: "Think about how you would answer differently or add to the answer."

5. Team Discussion: As a team, discuss other possible answers or reactions to the answer given.

Question Commander

1. Ask the Question: Question Commander reads the question to the team.

2. Think Time: "Think of your best answer."

3. Answer the Question: The Question Commander selects a teammate to answer the question.

4. Think Time: "Think about how you would answer differently or add to the answer."

5. Team Discussion: As a team, discuss other possible answers or reactions to the answer given.

Question Commander

1. Ask the Question: Question Commander reads the question to the team.

2. Think Time: "Think of your best answer."

3. Answer the Question: The Question Commander selects a teammate to answer the question.

4. Think Time: "Think about how you would answer differently or add to the answer."

5. Team Discussion: As a team, discuss other possible answers or reactions to the answer given.

Fan-N-Pick

In a team of four, Student One fans out the question cards, and says, "Pick a card, any card!" Student Two picks a card and reads the question out loud to teammates. After five seconds of think time, Student Three gives his or her answer. After another five seconds of think time, Student Four paraphrases, praises, or adds to the answer given. Students rotate roles for each new round.

Spin-N-Think™

Spin-N-Think spinners are available from Kagan to lead teams through the steps of higher-level thinking. Students spin the Spin-N-Think™ spinner to select a student at each stage of the questioning to: 1) ask the question, 2) answer the question, 3) paraphrase and praise the answer, 4) augment the answer, and 5) discuss the question or answer. The Spin-N-Think™ game makes higher-level thinking more fun, and holds students accountable because they are often called upon, but never know when their number will come up.

Three-Step Interview

After the question is read to the team, students pair up. The first step is an interview in which one student interviews the other about the question. In the second step, students remain with their partner but switch roles: The interviewer becomes the interviewee. In the third step, the pairs come back together and each student in turn presents to the team what their partner shared. Three-Step Interview is strong for individual accountability, active listening, and paraphrasing skills.

Team Discussion

Team Discussion is an easy and informal way of processing the questions: Students read a question and then throw it open for discussion. Team Discussion, however, does not ensure that there is individual accountability or equal participation.

Think-Pair-Square

One student reads a question out loud to teammates. Partners on the same side of the table then pair up to discuss the question and their answers. Then, all four students come together for an open discussion about the question.

Question-Write-RoundRobin

Students take turns asking the team the question. After each question is asked, each student writes his or her ideas on a piece of paper. After students have finished writing, in turn they share their ideas. This format creates strong individual accountability because each student is expected to develop and share an answer for every question.

Mix-Pair-Discuss

Each student gets a different Question Card. For 16 to 32 students, use two sets of questions. In this case, some students may have the same question which is OK. Students get out of their seats and mix around the classroom. They pair up with a partner. One partner reads his or her Question Card and the other answers. Then they switch roles. When done they trade cards and find a new partner. The process is repeated for a predetermined amount of time. The rule is students cannot pair up with the same partner twice. Students may get the same questions twice or more, but each time it is with a new partner. This strategy is a fun, energizing way to ask and answer questions.

Think-Pair-Share

Think-Pair-Share is teacher-directed. The teacher asks the question, then gives students think time. Students then pair up to share their thoughts about the question. After the pair discussion, one student is called on to share with the class what was shared in his or her pair. Think-Pair-Share does not provide as much active participation for students as Think-Pair-Square because only one student is called upon at a time, but is a nice way to do whole-class sharing.

Inside-Outside Circle

Each student gets a Question Card. Half of the students form a circle facing out. The other half forms a circle around the inside circle; each student in the outside circle faces one student in the inside circle. Students in the outside circle ask inside circle students a question. After the inside circle students answer the question, students switch roles questioning and answering. After both have asked and answered a question,

they each praise theother's answers and then hold up a hand indicating they are finished. When most students have a hand up, have students trade cards with their partner and rotate to a new partner. To rotate, tell the outside circle to move to the left. This format is a lively and enjoyable way to ask questions and have students listen to the thinking of many classmates.

Question & Answer

This might sound familiar: Instead of giving students the Question Cards, the teacher asks the questions and calls on one student at a time to answer. This traditional format eliminates simultaneous, cooperative interaction, but may be good for introducing younger students to higher-level questions.

Numbered Heads Together

Students number off in their teams so that every student has a number. The teacher asks a question. Students put their "heads together" to discuss the question. The teacher then calls on a number and selects a student with that number to share what his or her team discussed.

pair activity #1

RallyRobin

Each pair gets a set of Question Cards. Student A in the pair reads the question out loud to his or her partner. Student B answers. Partners take turns asking and answering each question.

Pair Discussion

Partners take turns asking the question. The pair then discusses the answer together. Unlike RallyRobin, students discuss the answer. Both students contribute to answering and to discussing each other's ideas.

Question-Write-Share-Discuss

One partner reads the Question Card out loud to his or her teammate. Both students write down their ideas. Partners take turns sharing what they wrote. Partners discuss how their ideas are similar and different.

Journal Writing

Students pick one Question Card and make a journal entry or use the question as the prompt for an essay or creative writing. Have students share their writing with a partner or in turn with teammates.

Independent Answers

Students each get their own set of Questions Cards. Pairs or teams can share a set of questions, or the questions can be written on the board or put on the overhead projector. Students work by themselves to answer the questions on a separate sheet of paper. When done, students can compare their answers with a partner, teammates, or the whole class.

Center Ideas

1. Question Card Center

At one center, have the Question Cards and a Spin-N-Think™ spinner, Question Commander instruction card, or Fan-N-Pick instructions. Students lead themselves through the thinking questions. For individual accountability, have each student record their own answer for each question.

2. Journal Writing Center

At a second center, have a Journal Writing activity page for each student. Students can discuss the question with others at their center, then write their own journal entry. After everyone is done writing, students share what they wrote with other students at their center.

3. Question Starters Center

At a third center, have a Question Starters page. Split the students at the center into two groups. Have both groups create thinking questions using the Question Starters activity page. When the groups are done writing their questions, they trade questions with the other group at their center. When done answering each other's questions, two groups pair up to compare their answers.

Higher-Level Thinking Questions for Secondary Mathematics
Kagan Publishing • 1 (800) 933-2667 • www.KaganOnline.com

Problem Solving

higher-level thinking questions

"An expert problem solver must be endowed with two incompatible qualities, a restless imagination and a patient pertinacity.

— Howard W. Eves

Higher-Level Thinking Questions for Secondary Mathematics
Kagan Publishing • 1 (800) 933-2667 • www.KaganOnline.com

Problem Solving
Question Cards

Problem Solving

1 What are some instances in which you had to solve a multistep problem in real life? Describe the problem and the steps you used to solve it.

Problem Solving

2 What are some things you would suggest that might improve the teaching of problem solving?

Problem Solving

3 In what ways can pictures, objects, or symbols lead you to the solution of a problem?

Problem Solving

4 If a problem can be solved using more than one strategy, what are some questions you ask yourself to decide which strategy to use?

Problem Solving
Question Cards

5 What connections can you make between math concepts or skills and problem solving?

6 In your own words and in your own life, what is problem solving?

7 What are some things common to all solvable problems?

8 When you are "stuck" in the middle of solving a problem, what are some things you think about to help you get "unstuck"?

Higher-Level Thinking Questions for Secondary Mathematics
Kagan Publishing • 1 (800) 933-2667 • www.KaganOnline.com

Problem Solving
Question Cards

Problem Solving

9 How would you complete the analogy, "Problem solving in math is to learning math skills as playing a _____ is to ___"? Explain your answer.

Problem Solving

10 In your opinion, which step of the problem-solving process (Understand, Plan, Solve, Check) could most easily be eliminated? Justify your answer.

Problem Solving

11 What are some ways you would approach solving a problem in which a plan is not immediately obvious?

Problem Solving

12 Process and strategies are used to solve problems in mathematics. What are some processes or strategies used to solve problems in reading, science, and social studies? Use specific examples to justify your answer.

Problem Solving
Question Cards

Problem Solving

13 The problem-solving process has four steps that are performed in a specific sequence: Understand, Plan, Solve, Check. What are some things you do on a regular basis that have a definite and fixed sequence?

Problem Solving

14 If you were a teacher, what kinds of problems might you encounter during your normal workday? How would you solve them?

Problem Solving

15 What are some of the factors that affect how you would solve a problem?

Problem Solving

16 You are inventing a new problem solving process. What are its steps? What is the purpose of each step?

Higher-Level Thinking Questions for Secondary Mathematics
Kagan Publishing • 1 (800) 933-2667 • www.KaganOnline.com

Problem Solving

Journal Writing Question

Write your response to the question below.
Be ready to share your response.

What are some instances in which you had to solve a multistep problem in real life? Describe the problem and the steps you used to solve it.

_____
_____
_____
_____
_____
_____
_____
_____
_____
_____
_____
_____
_____
_____
_____
_____
_____
_____

Problem Solving
Question Starters

Use the question starters below to create complete questions.
Send your questions to a partner or to another team to answer.

1. Why is problem solving

2. How can a problem be solved if

3. When solving a problem,

4. How can patterns in problems

5. How are all problems

6. What information in a problem is

7. If one condition of the problem changes, how

8. How can checking a problem's solution

Higher-Level Thinking Questions for Secondary Mathematics
Kagan Publishing • 1 (800) 933-2667 • www.KaganOnline.com

Number Patterns
and Relationships

higher-level thinking questions

Mathematicians have tried in vain to this day to discover some order in the sequence of prime numbers, and we have reason to believe that it is a mystery into which the human mind will never penetrate.

— Leonhard Euler

Higher-Level Thinking Questions for Secondary Mathematics
Kagan Publishing • 1 (800) 933-2667 • www.KaganOnline.com

Number Patterns and Relationships
Question Cards

Number Patterns and Relationships

1 What are some ways number properties, such as commutative, associative, and distributive, describe the relationships among numbers?

Number Patterns and Relationships

2 How would you complete the analogy, "The commutative property is to peanut butter and jelly as the associative property is to ___"? Describe your thinking.

Number Patterns and Relationships

3 How can you relate the associative property to a similar procedure in your everyday life? Describe the similarities.

Number Patterns and Relationships

4 You are explaining number patterns and relationships to an elementary school student. How would you begin? What would you say?

---

5 What are some number patterns that you see on a regular basis? Why do you think they are used so regularly?

---

6 What observations can be made about number patterns and relationships in expressions with exponents and powers?

---

7 The Math Committee has elected you to rename the term exponent. What will you call it? Give reasons for the new name.

---

8 "Every even number can be written as the sum of two primes." What are some ways you could prove or disprove this statement?

Higher-Level Thinking Questions for Secondary Mathematics
Kagan Publishing • 1 (800) 933-2667 • www.KaganOnline.com

Number Patterns and Relationships
Question Cards

Number Patterns and Relationships

9 What experiments could you design that would test divisibility by 24, 30, or any other two-digit number? How do these experiments utilize number patterns?

Number Patterns and Relationships

10 "All number relationships are described or defined in properties." Do you agree or disagree with this statement? Explain.

Number Patterns and Relationships

11 An order of operations tells you the sequence to perform operations. Does your favorite hobby have an order of operations? Describe the order and what would happen if you mixed up the order.

Number Patterns and Relationships

12 The number 64 can be represented as the product of 16 and 4. What are some other ways to represent 64? What is the relationship among these representations?

Number Patterns and Relationships
Question Cards

Number Patterns and Relationships

13 What are some ways you could use drawings or diagrams to demonstrate the relationship between prime and composite numbers?

Number Patterns and Relationships

14 What are some ways that number relationships and patterns affect security?

Number Patterns and Relationships

15 How are number relationships similar to relationships among people? How are they different?

Number Patterns and Relationships

16 You have been asked to find all the ways 124 CDs could be packaged. What are some ways you could use number relationships to solve the problem?

Higher-Level Thinking Questions for Secondary Mathematics
Kagan Publishing • 1 (800) 933-2667 • www.KaganOnline.com

Number Patterns and Relationships

Journal Writing Question

Write your response to the question below.
Be ready to share your response.

What are some ways that number relationships and patterns affect security?

Number Patterns and Relationships

Question Starters

Use the question starters below to create complete questions.
Send your questions to a partner or to another team to answer.

1. How are number properties analogous to properties of

2. Can you use number relationships to

3. How would operating (working) with numbers change if

4. How are relationships among numbers like relationships with

5. How are numbers used to describe

6. How do number relationships

7. What do exponents and powers

8. How can number relationships be modeled using

Higher-Level Thinking Questions for Secondary Mathematics
Kagan Publishing • 1 (800) 933-2667 • www.KaganOnline.com

Whole Number and Decimal Place Value

higher-level thinking questions

> # "I advise my students to listen carefully the moment they decide to take no more mathematics courses. They might be able to hear the sound of closing doors."
>
> — James Caballero

Higher-Level Thinking Questions for Secondary Mathematics
Kagan Publishing • 1 (800) 933-2667 • www.KaganOnline.com

Whole Number and Decimal Place Value
Question Cards

Whole Number and Decimal Place Value

1 To safeguard against errors, check amounts are written in both standard form and word form. What are some other real-world examples of values being represented in more than one way?

Whole Number and Decimal Place Value

2 The Dewey Decimal System is used to systematize books in a library. In what other situations are whole numbers and decimals used for organizational purposes?

Whole Number and Decimal Place Value

3 You want to analyze the statistics for your favorite sport. Which decimal concepts would be essential to understand? Explain your thinking.

Whole Number and Decimal Place Value

4 What are some of the similarities and differences of terminating and repeating decimals?

Whole Number and Decimal Place Value

5 What makes each decimal value unique?

Whole Number and Decimal Place Value

6 What are some ways you can represent a number, such as 45.37, to someone who has not learned about decimals? Describe what you would say and do.

Whole Number and Decimal Place Value

7 What real-life concepts or items might you use to represent 0.5?

Whole Number and Decimal Place Value

8 What might happen if there were no decimals? Describe a week with no method for communicating decimal parts of a whole.

Higher-Level Thinking Questions for Secondary Mathematics
Kagan Publishing • 1 (800) 933-2667 • www.KaganOnline.com

Whole Number and Decimal Place Value
Question Cards

Whole Number and Decimal Place Value

9 What are some ways that decimal equivalences can be modeled using coins? Explain your thinking with specific examples.

Whole Number and Decimal Place Value

10 Whole number and decimal values can be expressed using the word form, standard form, or with expanded notation. Which do you believe is easiest to use in describing decimal values to others? Justify your reasoning.

Whole Number and Decimal Place Value

11 What are some strategies you use to compare and order decimals?

Whole Number and Decimal Place Value

12 A newspaper article certifies that a large school system's budget for one year is 1.4 billion dollars. What are some reasons that the newspaper may choose to write the amount as a decimal rather than its whole number equivalent?

Whole Number and Decimal Place Value
Question Cards

13 When reading a number, such as 34.5, the ten's place is two places to the left of the decimal point, while the tenth's place is to the immediate right of the decimal point? Why does this make sense?

14 The value of zero can be written as 0, 0.0, 0.00, and so forth. What are some reasons why zero is represented any way other than 0? Give examples to support your response.

15 Decimals are just as important as whole numbers. Do you agree or disagree with this statement? Why?

16 Whole number and decimal place value extend a base ten pattern. What are some patterns in your life that could be extended? In what ways would you extend them?

50

Higher-Level Thinking Questions for Secondary Mathematics
Kagan Publishing • 1 (800) 933-2667 • www.KaganOnline.com

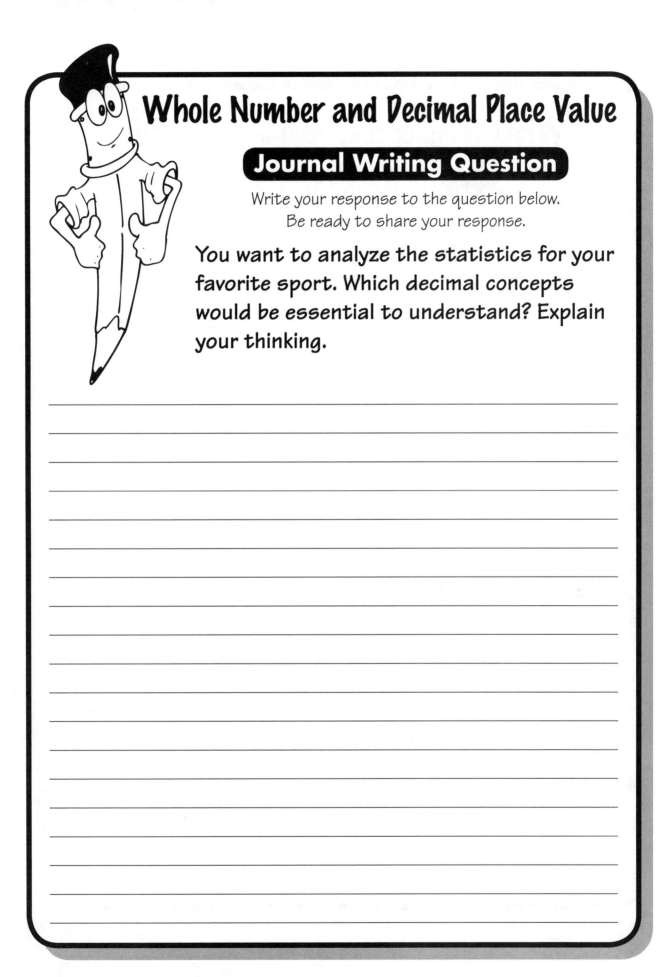

Whole Number and Decimal Place Value

Journal Writing Question

Write your response to the question below.
Be ready to share your response.

You want to analyze the statistics for your favorite sport. Which decimal concepts would be essential to understand? Explain your thinking.

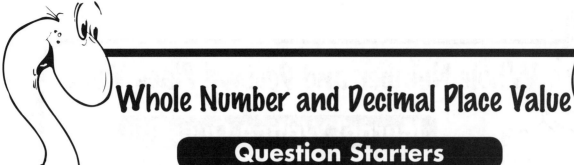

Whole Number and Decimal Place Value

Question Starters

Use the question starters below to create complete questions.
Send your questions to a partner or to another team to answer.

1. Where or when have you seen decimals represented as

2. How do decimal concepts apply to

3. How are decimals and whole numbers

4. Why are whole numbers and decimals used to

5. Why is it best to use decimals when

6. How would writing decimals be different if

7. What if whole numbers and decimal

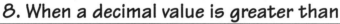

8. When a decimal value is greater than

Higher-Level Thinking Questions for Secondary Mathematics
Kagan Publishing • 1 (800) 933-2667 • www.KaganOnline.com

Decimal
Operations

higher-level thinking questions

"The truth of the matter is that, though mathematics truth may be beauty, it can be only glimpsed after much hard thinking. Mathematics is difficult for many human minds to grasp because of its hierarchical structure: one thing builds on another and depends on it.

— M. Holt and D. T. E. Marjoram

Decimal Operations
Question Cards

1 What are some ways that balancing a checkbook is an application of decimal concepts? Describe as many ways as you can.

2 There are many ways to calculate a tip and include it in the total amount. Most restaurant tips are between 15–20 percent of the total cost of the meal. Which method would you find most comfortable for calculating the tip in that range? Explain your thinking.

3 Roman Numerals can be used to represent whole numbers. The letters I, V, and X represent 1, 5, and 10, respectively. The number 26 is written as XXVI. Do you think Roman Numerals could be used to represent decimal amounts? If not, why not? If so, how might twenty-six hundredths be written?

4 What are some four-number sets that produce a combined sum of 10.2? Explain the process you used to arrive at a solution.

Decimal Operations
Question Cards

Decimal Operations

5 You want to mentally compute the savings of an item while standing in the store. What decimal number patterns, or "tricks," might you look for to help you make the mental calculations easier?

Decimal Operations

6 Medical procedures and prescriptions are measured in decimal metric units. What are some reasons that decimal operations are critical for you to know as a result of this fact?

Decimal Operations

7 You want to analyze your monthly earnings and expenses at the end of the year. What questions might you ask yourself? What are some ways you could organize the decimal (money) amounts to make your computation easier?

Decimal Operations

8 At a store, you find that a product comes in small, medium, and large boxes. What combination of concepts or skills would you use in order to identify the size of box that is the best buy?

Higher-Level Thinking Questions for Secondary Mathematics
Kagan Publishing • 1 (800) 933-2667 • www.KaganOnline.com

Decimal Operations
Question Cards

Decimal Operations

9 A student is confused about whether to multiply or divide when converting metric units. Play the role of the teacher. What would you say to help the student make sense of the conversion processes?

Decimal Operations

10 What are some ways you can simplify the process of multiplying two decimals?

Decimal Operations

11 Scientific notation represents values using decimals and powers of ten. For example, the Sun is 93,000,000, or 9.3×10^7 miles away. What are some criteria to decide whether or not to use scientific notation to express a value? Use examples to explain your thinking.

Decimal Operations

12 Three strategies for estimating decimal products involve using compatible numbers, rounding, and the distributive property. What do you feel are the pluses and minuses of using each of these three strategies?

Decimal Operations
Question Cards

13 Each state has its own rate of sales tax. If your state could generate more revenue for social prorams with a higher tax rate, would you favor the tax increase? Explain your reasoning.

14 How would you complete the analogy, "Decimals are to money as _____ is to _____"? Explain your reasoning.

15 What are some general statements you can make about predicting the relative size of a quotient when a decimal is divided by another decimal?

16 Suppose you chose to eliminate the communication of all metric units of length except one. This would mean that you would use decimal operations to convert all measurements to one particular unit. Which unit would you choose and why?

Higher-Level Thinking Questions for Secondary Mathematics
Kagan Publishing • 1 (800) 933-2667 • www.KaganOnline.com

Decimal Operations
Journal Writing Question

Write your response to the question below.
Be ready to share your response.

Three strategies for estimating decimal products involve using compatible numbers, rounding, and the distributive property. What do you feel are the pluses and minuses of using each of these three strategies?

Decimal Operations
Question Starters

*Use the question starters below to create complete questions.
Send your questions to a partner or to another team to answer.*

1. When multiplying two decimal amounts,

2. How is finding the average of several decimal amounts

3. How can working with decimals help to

4. Without decimals,

5. How would writing decimals be different if

6. How can you predict the decimal product or quotient when

7. Why is subtracting decimals similar to

8. What decimal amounts

Higher-Level Thinking Questions for Secondary Mathematics
Kagan Publishing • 1 (800) 933-2667 • www.KaganOnline.com

Fraction
Concepts

higher-level thinking questions

"The whole is more than the sum of its parts."

— Aristotle

Higher-Level Thinking Questions for Secondary Mathematics
Kagan Publishing • 1 (800) 933-2667 • www.KaganOnline.com

Fraction Concepts
Question Cards

1 What are some ways to classify fractions? Which fractions might be in each classification?

2 It is necessary to find equivalent fractions for many mathematical procedures. What are some real-life instances in which you might need to find equivalent fractions?

3 Architects use fractions when they create blueprints for homes. What are some other professions and careers that require fluency in fractions? Explain your thinking.

4 Equivalent fractions have the same value, but different names. What are some other examples of things that have the same meaning or value but are expressed in different ways?

Fraction Concepts
Question Cards

5 You are preparing a report in which you must share statistics in fractional form. Would it be a better decision to simplify the fractions or to keep them in their original form? Explain the reasons for your opinion.

6 What are some ways you could help a classmate make sense of equivalent fractions using words, pictures, or objects?

7 What are some questions related to fractions that you ask yourself when measuring ingredients and following a recipe?

8 What is your strategy for ordering fractions? Describe what you do and why you do it.

Higher-Level Thinking Questions for Secondary Mathematics
Kagan Publishing • 1 (800) 933-2667 • www.KaganOnline.com

Fraction Concepts

9 "Fractions define division." Do you agree or disagree with this statement? Give reasons for your response.

Fraction Concepts

10 Items from shoes to tennis racquets to cakes are sold in fractional sizes. Cakes and pies may be sold in fractional parts. What are some items sold in wholes only that you would like to see sold in fractional sizes or parts? Explain your thinking, and offer strategies for sizing or pricing.

Fraction Concepts

11 In customary measurement, one quart is 1/4 of a gallon, a foot is 1/3 of a yard, and an ounce is 1/16 of a pound. What are the similarities and differences in the processes of changing measurements from one customary unit to another?

Fraction Concepts

12 A cubit is the "distance between a person's elbow to the end of the middle finger." A span is the "distance between the end of a person's thumb and pinky fingers when the hand is outstretched." How would you investigate whether or not there is a fractional relationship between a person's cubit, span, and height? In what ways would you analyze your results?

Fraction Concepts
Question Cards

Fraction Concepts

13 A party planner finds that packages of cups, plates, and napkins are sold in different quantities. She wants to have the same number of each item. What advice would you give the party planner? Explain your thinking.

Fraction Concepts

14 Sometimes you need an equivalent fraction with a smaller denominator while other times an equivalent fraction with a larger denominator is needed. What are some instances that each is necessary?

Fraction Concepts

15 While customary measurement utilizes fractions, metric measurement makes use of decimals. If you lived in a country that used only metric measurement, would you need to learn about fractions? Why or why not?

Fraction Concepts

16 Fractions can be defined as "part of a whole or part of a group." In what ways are you a fractional part of a whole or part of a group? Explain.

Higher-Level Thinking Questions for Secondary Mathematics
Kagan Publishing • 1 (800) 933-2667 • www.KaganOnline.com

Fraction Concepts
Journal Writing Question

Write your response to the question below.
Be ready to share your response.

Items from shoes to tennis racquets to cakes are sold in fractional sizes. Cakes and pies may be sold in fractional parts. What are some items sold in wholes only that you would like to see sold in fractional sizes or parts? Explain your thinking, and offer strategies for sizing or pricing.

Fraction Concepts

Question Starters

Use the question starters below to create complete questions.
Send your questions to a partner or to another team to answer.

1. Why are fractions

2. How is the relationship between fractions and

3. When is the fractional part of a whole or a group

4. If there were no fractions, how would

5. How might fractions be used

6. In your opinion, what fraction

7. How does the denominator of a fraction affect

8. How can improper fractions

Higher-Level Thinking Questions for Secondary Mathematics
Kagan Publishing • 1 (800) 933-2667 • www.KaganOnline.com

Fraction Addition and Subtraction

higher-level thinking questions

"There are many questions which fools can ask that wise men cannot answer.

— George Pólya

Fraction Addition and Subtraction
Question Cards

Fraction Addition and Subtraction

1 Adding and subtracting mixed numbers requires knowledge of a combination of math skills and concepts. Name one accomplishment that you are proud of. What skills and knowledge were required for you to accomplish this goal?

Fraction Addition and Subtraction

2 When might it be preferable to round fractions or mixed numbers to the nearest whole number?

Fraction Addition and Subtraction

3 You found your grandmother's family recipes. Each recipe lists the ingredients, but doesn't indicate the serving size. To calculate the serving size for the recipe, could you add the ingredients listed or are there other things to consider?

Fraction Addition and Subtraction

4 What are the advantages and disadvantages of adding and subtracting fractions or mixed numbers, rather than converting them to decimals and finding their sums and differences?

Fraction Addition and Subtraction
Question Cards

Fraction Addition and Subtraction

5 A carpenter is using a blueprint to purchase lumber so she can frame a house. What are some ways she must use fractions and mixed numbers?

Fraction Addition and Subtraction

6 What are some questions you ask yourself when subtracting a mixed number from a whole number?

Fraction Addition and Subtraction

7 What are some strategies for rounding fractions and mixed numbers?

Fraction Addition and Subtraction

8 As the Chief Financial Officer of a company, you closely follow your company's stock prices and that of your competitors. How will your knowledge of adding and subtracting fractions help you?

Higher-Level Thinking Questions for Secondary Mathematics
Kagan Publishing • 1 (800) 933-2667 • www.KaganOnline.com

Fraction Addition and Subtraction
Question Cards

Fraction Addition and Subtraction

9 A classmate says, "To add 1/2 and 1/2, add the numerators and then add the denominators." Do you agree or disagree with this as a general rule for adding fractions? Explain.

Fraction Addition and Subtraction

10 How are the procedures for adding mixed numbers and subtracting mixed numbers alike? How are they different?

Fraction Addition and Subtraction

11 What are some fraction pairs that can be combined to produce a whole number? What patterns do you see in these fraction pairs?

Fraction Addition and Subtraction

12 An architect knows that there must be at least 2 2/3 feet clearance beside a bathtub for maximum comfort. What are some other observations an architect must make when designing a home?

Fraction Addition and Subtraction
Question Cards

Fraction Addition and Subtraction

13 What are some instances that an exact sum or difference is needed, rather than an estimate?

Fraction Addition and Subtraction

14 There is a sequence of steps used to add or subtract fractions. What parts of the sequence could be changed, and what parts must remain the same? Explain your reasoning using examples.

Fraction Addition and Subtraction

15 To design an outfit, you need 5 1/2 yards of fabric, but you only have 2 2/3 yards. How do you know how much fabric to buy, and what special considerations should you take into account?

Fraction Addition and Subtraction

16 How would you complete the analogy, "Subtracting fractions is to subtracting mixed numbers as _____ is to _____"?

Higher-Level Thinking Questions for Secondary Mathematics
Kagan Publishing • 1 (800) 933-2667 • www.KaganOnline.com

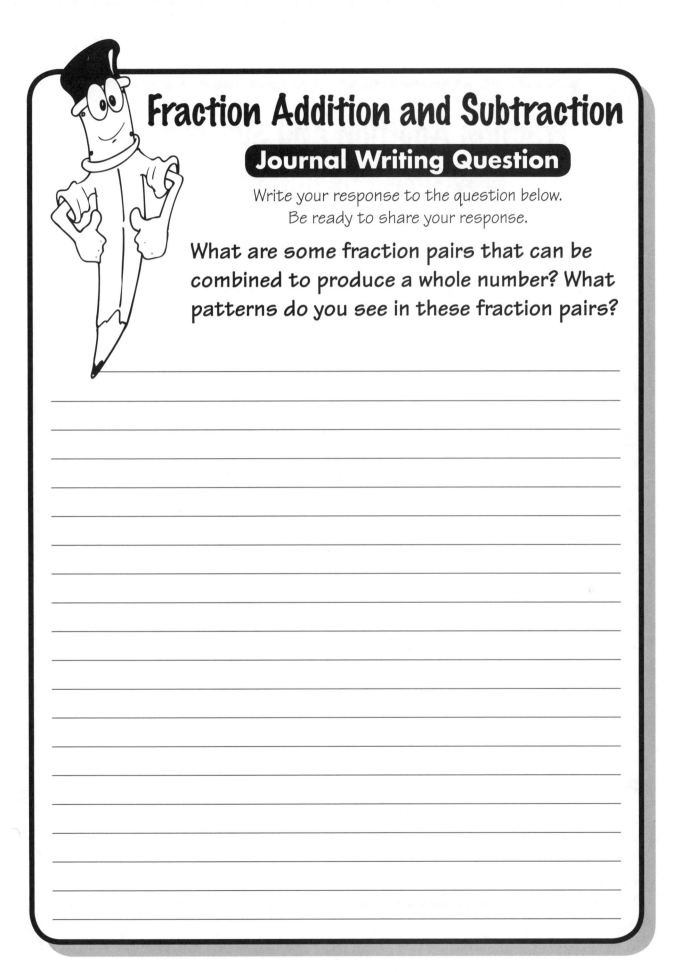

Fraction Addition and Subtraction

Journal Writing Question

Write your response to the question below.
Be ready to share your response.

What are some fraction pairs that can be combined to produce a whole number? What patterns do you see in these fraction pairs?

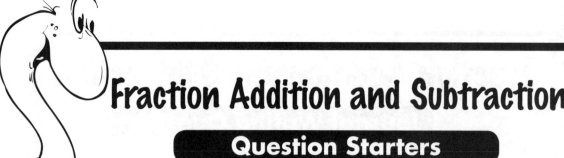

Fraction Addition and Subtraction

Question Starters

Use the question starters below to create complete questions.
Send your questions to a partner or to another team to answer.

1. When is the sum of two fractions

_____

_____

2. What might happen when three or more fractions

_____

_____

3. When comparing and ordering fractions, why

_____

_____

4. When subtracting mixed numbers, when is it necessary to

_____

_____

5. How can adding fractions

_____

_____

6. What sequence may be used when two fractions

_____

_____

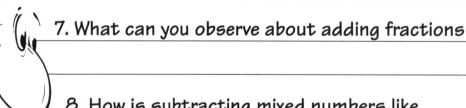

7. What can you observe about adding fractions

_____

8. How is subtracting mixed numbers like

_____

_____

Higher-Level Thinking Questions for Secondary Mathematics

Kagan Publishing • 1 (800) 933-2667 • www.KaganOnline.com

Fraction Multiplication and Division

higher-level thinking questions

"Mathematics seems to endow one with something like a new sense.

— Charles Darwin

Fraction Multiplication and Division
Question Cards

Fraction Multiplication and Division

1 Before you begin to find the product of two mixed numbers, you must first change mixed numbers into improper fractions. What are some everyday tasks in which you must do something before you can begin the process you wish to do?

Fraction Multiplication and Division

2 What are some things you think about when estimating the product of two mixed numbers?

Fraction Multiplication and Division

3 In the division of fractions and mixed numbers, sometimes the quotient is the greatest value. What about the relative sizes of the divisor, dividend, and quotient causes this to happen?

Fraction Multiplication and Division

4 What are the similarities and differences in the greatest common factor and the least common multiple?

Fraction Multiplication and Division
Question Cards

Fraction Multiplication and Division

5 In what ways does a chef multiply and divide fractions when he alters a basic recipe to feed more people or change a serving size?

Fraction Multiplication and Division

6 What connections can you make between multiplication and repeated addition? How might you find the product of two mixed numbers using repeated addition?

Fraction Multiplication and Division

7 What observations do you make about a problem situation that help you determine whether to solve by multiplying or dividing fractions?

Fraction Multiplication and Division

8 A company president multiplies her sales force's yearly sales by a mixed number to determine their sales goals for the next year. What are some things you know about the sales goals?

Higher-Level Thinking Questions for Secondary Mathematics
Kagan Publishing • 1 (800) 933-2667 • www.KaganOnline.com

Fraction Multiplication and Division
Question Cards

Fraction Multiplication and Division

9 What are some practical applications of multiplying or dividing mixed numbers that you might use in everyday life?

Fraction Multiplication and Division

10 You have invented a method for mentally computing the product of two fractions. What are some ways you can explain how your invention works to classmates so that they can use it? Use examples in your explanation.

Fraction Multiplication and Division

11 What are some ways you could use drawings, diagrams, or objects to illustrate the division of two mixed numbers?

Fraction Multiplication and Division

12 What are some meanings of the word divide? How do those meanings relate to the mathematical process of division?

Fraction Multiplication and Division
Question Cards

13 What general rule can you make about fraction pairs whose products are smaller than either factor?

14 Storeowners increase the price of a garment by a "profit" factor between 3/4 and 2 1/2 before selling the garment to you. What are some conditions that might affect the owner's decision about the size of the profit factor?

15 What are some ways you can interpret the remainder when dividing two fractions or mixed numbers? Give examples to support your answer.

16 What is the opposite of dividing two mixed numbers? Explain your thinking.

Higher-Level Thinking Questions for Secondary Mathematics
Kagan Publishing • 1 (800) 933-2667 • www.KaganOnline.com

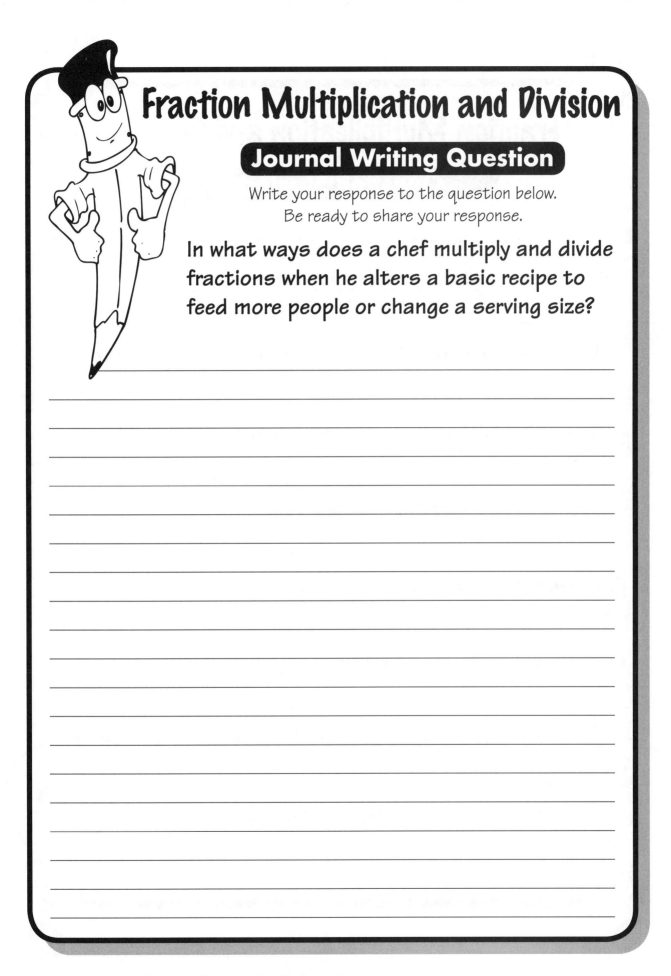

Fraction Multiplication and Division

Journal Writing Question

Write your response to the question below.
Be ready to share your response.

In what ways **does a chef multiply and divide**
fractions when he alters a basic recipe to
feed more people or change a serving size?

Fraction Multiplication and Division

Question Starters

Use the question starters below to create complete questions.
Send your questions to a partner or to another team to answer.

1. What happens when you multiply a mixed number by

2. Why is the size of the divisor in a division problem related to

3. What is the relationship between the mixed number factors and

4. When dividing a mixed number, how does

5. What patterns in fractions help to predict

6. How could you represent mixed number multiplication using

7. What general rules about multiplication and division

8. How can multiplying mixed numbers help to

Higher-Level Thinking Questions for Secondary Mathematics
Kagan Publishing • 1 (800) 933-2667 • www.KaganOnline.com

Percents

higher-level thinking questions

"The mathematician has reached the highest rung on the ladder of human thought.

— Havelock Ellis

Higher-Level Thinking Questions for Secondary Mathematics
Kagan Publishing • 1 (800) 933-2667 • www.KaganOnline.com

Percents
Question Cards

Percents

1 A football player's salary during the second year of play may be 175% of his first year's salary. What are some conclusions that you can draw about the player's salary?

Percents

2 A store has a sale in which everything is 25% off the marked price. What are some ways you can mentally calculate the sale price of an item you like? How might you check your computation to make sure it is reasonable?

Percents

3 In order to find the percent one number is of another, what are some questions you might ask yourself?

Percents

4 A photograph developed from 35 mm film is 428% larger than the negative. What are some other real-life examples of percents that are greater than 100%?

Percents
Question Cards

Percents

5 The price of a computer game that you have been thinking about buying is reduced by 65%. What might have caused the company to reduce the price so drastically?

Percents

6 What are some ways you could solve a percentage problem other than using percents?

Percents

7 A bank pays interest on money you deposit and charges interest on money you borrow. The interest rate on deposited money is lower than the interest on borrowed money. What are some things that would happen if the reverse were true?

Percents

8 Test scores are often recorded in percentages. What general rules or strategies would you use to find the score percentage of an assessment when the number of items is not a factor or a multiple of 100?

Higher-Level Thinking Questions for Secondary Mathematics
Kagan Publishing • 1 (800) 933-2667 • www.KaganOnline.com

Percents
Question Cards

Percents

9 Polling a random sample of a population provides data without counting the whole population. If a random sample reveals that 75% of your school population prefers cats to dogs, what accurate and inaccurate conclusions might you draw?

Percents

10 Pollsters may predict election results with a leeway of about 3%. What are some of the advantages and disadvantages to these predictions? What is your opinion of publicly announcing these predictions on Election Day?

Percents

11 A baseball player's batting average is the percent of times at bat that result in hits. What are some other sports statistics that are expressed as percentages and why are they important to know?

Percents

12 What are some ways you can represent percents or percentages symbolically or visually? Explain.

Percents
Question Cards

Percents

13 The term percent means "per one hundred." You want to invent a similar ratio, comparing amounts to ten. What are some uses for your new ratio? What might be a suitable name?

Percents

14 Suppose that 47% of the students in your school are boys at the start of the school year. What might cause the percentage to increase or decrease over the course of the school year?

Percents

15 What are some situations in which 100% is the goal? What are some situations in which 0% is the goal? What makes these instances different from one another?

Percents

16 European waiters and waitresses may be paid a higher base salary than their American counterparts because they do not receive tips. If you were a waiter or waitress, would you prefer to work for tips or for a salary? Explain your reasoning.

Higher-Level Thinking Questions for Secondary Mathematics
Kagan Publishing • 1 (800) 933-2667 • www.KaganOnline.com

Percents
Journal Writing Question

Write your response to the question below.
Be ready to share your response.

Pollsters may predict election results with a leeway of about 3%. What are some of the advantages and disadvantages to these predictions? What is your opinion of publicly announcing these predictions on Election Day?

Percents

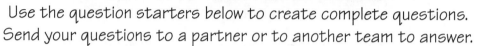

Question Starters

Use the question starters below to create complete questions.
Send your questions to a partner or to another team to answer.

1. Can every percentage be written

2. How are percentages used in

3. How is calculating a percent different from

4. In a life without percents,

5. What happens when the percentage of an amount is greater than

6. Is there a percentage for

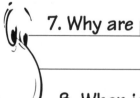
7. Why are percentages

8. When is the percent of a number

Higher-Level Thinking Questions for Secondary Mathematics
Kagan Publishing • 1 (800) 933-2667 • www.KaganOnline.com

Relating Fractions, Decimals, and Percents

higher-level thinking questions

Mathematics is the art of giving the same name to different things. [As opposed to the quotation: Poetry is the art of giving different names to the same thing.]

— Jules Henri Poincaré

Relating Fractions, Decimals, and Percents
Question Cards

Relating Fractions, Decimals, and Percents

1 The term benchmark means "standard" or a "point of reference." What are some benchmark percents? What are their decimal and fraction equivalents?

Relating Fractions, Decimals, and Percents

2 How would you complete the analogy, "Mixed numbers are to common fractions as large percents are to ___"?

Relating Fractions, Decimals, and Percents

3 You are e-mailing a trail mix recipe to a friend. Would you use fractions, decimals, or percents to communicate the relationship among the ingredients? Justify your opinion.

Relating Fractions, Decimals, and Percents

4 When should percents be used as opposed to fractions and decimals? Why?

Relating Fractions, Decimals, and Percents
Question Cards

Relating Fractions, Decimals, and Percents

5 What are some ways to visually represent the relationship among fractions, decimals, and percents?

Relating Fractions, Decimals, and Percents

6 Do you think there is a fraction equivalent to every decimal, and a decimal equivalent to every fraction?

Relating Fractions, Decimals, and Percents

7 What are some ways to test the equivalence of fraction, decimal, and percent values?

Relating Fractions, Decimals, and Percents

8 As a Chief Executive Officer (CEO) of a company, what types of percents would you like to know about your business? Explain your thinking.

Higher-Level Thinking Questions for Secondary Mathematics
Kagan Publishing • 1 (800) 933-2667 • www.KaganOnline.com

Relating Fractions, Decimals, and Percents

9 What distinguishes decimals from fractions and percents?

Relating Fractions, Decimals, and Percents

10 You have been told that fractions, decimals, or percents must be eliminated from everyday use. Which do you feel would be easiest to live without? Explain your reasoning.

Relating Fractions, Decimals, and Percents

11 What are some times in your daily life that it is necessary to convert between fractions, decimals, and percents?

Relating Fractions, Decimals, and Percents

12 You are having a sale at your clothing store. Do you advertise that you are discounting clearance items at 50% off, 1/2 off, or 0.5 off? Justify your decision.

Relating Fractions, Decimals, and Percents
Question Cards

Relating Fractions, Decimals, and Percents

13 You are making a number line between zero and one. You can include one fraction, one decimal, and one percent on the number line. Make your choices and defend your answer.

Relating Fractions, Decimals, and Percents

14 How are fractions, decimals, and percents alike? How are they different?

Relating Fractions, Decimals, and Percents

15 Individuals with higher incomes pay taxes at a higher percentage rate. Would you predict that the rich, middle class, or poor pay more taxes? Explain.

Relating Fractions, Decimals, and Percents

16 What are some questions you have about the relationship among fractions, decimals, and percents? How might you try to find the answers to your questions?

Higher-Level Thinking Questions for Secondary Mathematics
Kagan Publishing • 1 (800) 933-2667 • www.KaganOnline.com

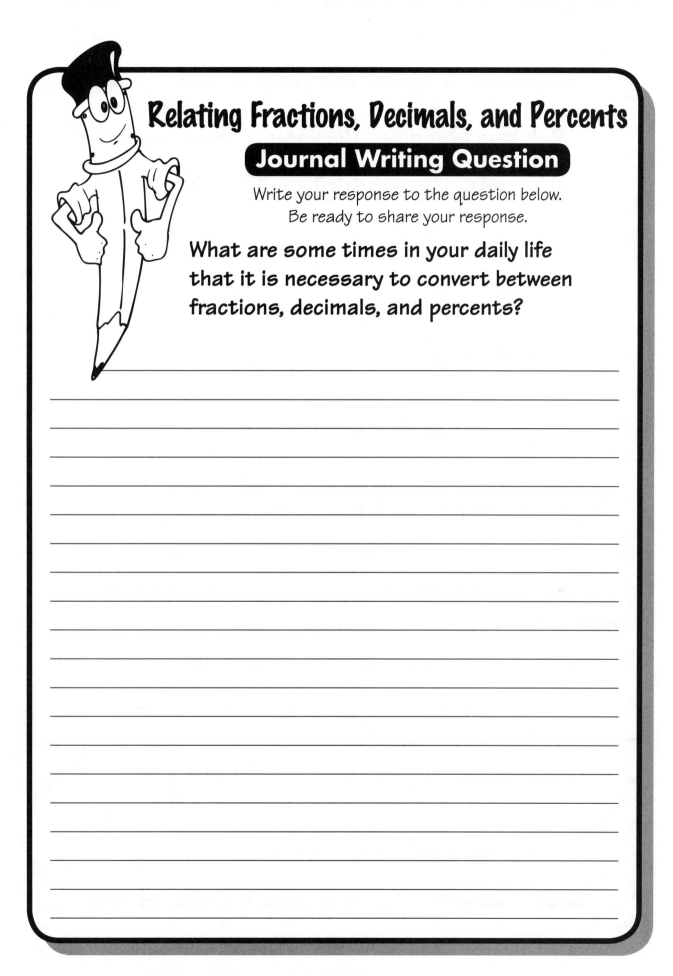

Relating Fractions, Decimals, and Percents

Journal Writing Question

Write your response to the question below.
Be ready to share your response.

What are some times in your daily life that it is necessary to convert between fractions, decimals, and percents?

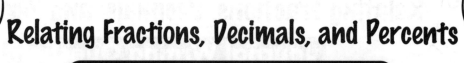

Relating Fractions, Decimals, and Percents

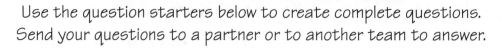

Question Starters

Use the question starters below to create complete questions.
Send your questions to a partner or to another team to answer.

1. To summarize data, percents

_____

2. How are numbers greater than 1 represented

_____

3. Why do fractions, decimals, and percents

_____

4. Construction workers use fractions when

_____

5. How would talking about fractions, decimals, and percents be

different if _____

6. What happens when a percent, fraction, or decimal

_____

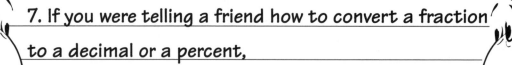

7. If you were telling a friend how to convert a fraction
to a decimal or a percent, _____

_____

8. A decimal such as 1.2 billion is used to represent

_____

Higher-Level Thinking Questions for Secondary Mathematics
Kagan Publishing • 1 (800) 933-2667 • www.KaganOnline.com

Ratios, Rates, and Proportions

higher-level thinking questions

Mathematics is not a careful march down a well-cleared highway, but a journey into a strange wilderness, where the explorers often get lost.

— W. S. Anglin

Ratios, Rates, and Proportions
Question Cards

Ratios, Rates, and Proportions

1 In what ways can ratios be used to analyze an athlete's performance?

Ratios, Rates, and Proportions

2 When are some times in your everyday life that you have used rate to determine the better buy? Describe the situation.

Ratios, Rates, and Proportions

3 In what ways are fractions and ratios alike? How are they different?

Ratios, Rates, and Proportions

4 The aspect ratio of a film is the relationship between the width of the film's image and the height. What issues do aspect ratios cause for TVs, movie screens, and computer monitors?

Ratios, Rates, and Proportions
Question Cards

Ratios, Rates, and Proportions

5 A ratio can be expressed as 2/6, 2 to 6, or 2:6. Do you think different expressions of the same ratio are more appropriate in different contexts? Describe when to use each.

Ratios, Rates, and Proportions

6 A rate is a "ratio that creates a relationship between two quantities with different kinds of units" like miles per hour (miles/hour). What are some other kinds of relationships between quantities you've seen and what characteristics do they share?

Ratios, Rates, and Proportions

7 What are some ways a builder can use what he or she knows from the blueprint to determine the appropriate amounts of materials to purchase?

Ratios, Rates, and Proportions

8 Population density is a "rate that compares the population per square mile." Would you rather live in an area with an unusually high or an unusually low population density? Explain your reasoning.

Higher-Level Thinking Questions for Secondary Mathematics
Kagan Publishing • 1 (800) 933-2667 • www.KaganOnline.com

Ratios, Rates, and Proportions
Question Cards

Ratios, Rates, and Proportions

9 A proportion is "an equation stating that two ratios are equivalent." Other than using cross products, what are some strategies for determining whether or not two ratios form a proportion? Use examples to explain.

Ratios, Rates, and Proportions

10 The rate of change shows how one quantity changes in relation to another quantity. How would you interpret a situation in which there was a decreasing rate of change? An increasing rate of change?

Ratios, Rates, and Proportions

11 Proportions are used to determine medicine dosage amounts. What are some conditions that cause the dosage amounts to differ among people?

Ratios, Rates, and Proportions

12 What are some times that proportions, ratios, or rates must be exact? What are some times that these relationships can be estimated?

Ratios, Rates, and Proportions

13 Maps are "scale drawings that show distances between two locations." What might help you decide whether to purchase a map with a relatively small scale or a relatively large scale? Explain your reasoning.

Ratios, Rates, and Proportions

14 Many countries use the metric system and use kilometers instead of miles to measure large distances. How would you convert the rate 65 miles per hour into kilometers per hour?

Ratios, Rates, and Proportions

15 In ratios, order matters. What is the difference between 3:5 and 5:3?

Ratios, Rates, and Proportions

16 Two differently priced items are reduced by the same amount. Would you predict that the percent of change for both items is the same? If not, which has the greater percentage of change? Explain.

Higher-Level Thinking Questions for Secondary Mathematics
Kagan Publishing • 1 (800) 933-2667 • www.KaganOnline.com

Ratios, Rates, and Proportions

Journal Writing Question

Write your response to the question below.
Be ready to share your response.

Two differently priced items are reduced by the same amount. Would you predict that the percent of change for both items is the same? If not, which has the greater percentage of change? Explain.

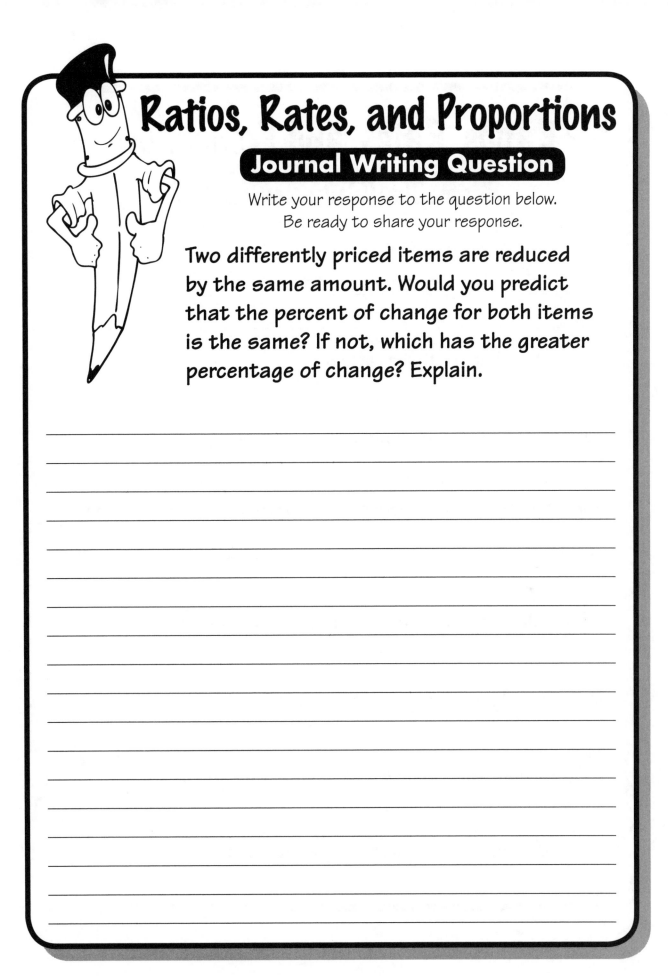

_____

_____

_____

_____

_____

_____

_____

_____

_____

_____

_____

_____

_____

_____

_____

_____

_____

Ratios, Rates, and Proportions

Question Starters

Use the question starters below to create complete questions.
Send your questions to a partner or to another team to answer.

1. What is the difference between rate and

2. What would happen if the ratio of

3. What patterns can you find between ratios and

4. If you had information about a rate, what predictions could you make about

5. When solving a proportion, what could you use for

6. How does the ratio of a number affect

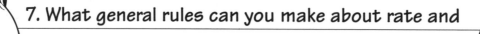

7. What general rules can you make about rate and

8. What is the opposite of a ratio that

Higher-Level Thinking Questions for Secondary Mathematics
Kagan Publishing • 1 (800) 933-2667 • www.KaganOnline.com

Statistics, Data Analysis, and Graphing

higher-level thinking questions

"The purpose of models is not to fit the data but to sharpen the questions.

— Samuel Karlin

Statistics, Data Analysis, and Graphing
Question Cards

1 Mean, median, and mode are three ways to help us understand data. What do each tell us?

2 If you were the president of the United States, which statistics would you be particularly interested in?

3 What conclusions can be drawn about a school or district by examining state or national assessment data?

4 What are some questions you ask yourself when deciding which graph would best represent data that you have collected?

Statistics, Data Analysis, and Graphing
Question Cards

Statistics, Data Analysis, and Graphing

5 Graphs are "visual representations of data that can be used to improve comprehension or understanding." When in your everyday life do you use a visual representation to help others understand a point you are making?

Statistics, Data Analysis, and Graphing

6 What kind of a graph would you invent? Tell its name, use, and how it would look.

Statistics, Data Analysis, and Graphing

7 What is the relationship between data and graphing?

Statistics, Data Analysis, and Graphing

8 What are some things you consider when you determine the scale of a graph?

Higher-Level Thinking Questions for Secondary Mathematics
Kagan Publishing • 1 (800) 933-2667 • www.KaganOnline.com

Statistics, Data Analysis, and Graphing
Question Cards

Statistics, Data Analysis, and Graphing

9 What preparations or plans should you make prior to collecting data on a topic?

Statistics, Data Analysis, and Graphing

10 Television shows are paid for by their commercial sponsors. What kinds of data do you think sponsors collect about the show before making a decision to advertise on it?

Statistics, Data Analysis, and Graphing

11 What are the similarities and differences between bar graphs and circle graphs?

Statistics, Data Analysis, and Graphing

12 Name one profession that requires great proficiency with graphs. How so?

Statistics, Data Analysis, and Graphing
Question Cards

Statistics, Data Analysis, and Graphing

13 What could a line graph tell you that you would likely have difficulty identifying looking at just the data?

Statistics, Data Analysis, and Graphing

14 What are some things you must consider when interpreting the data shown in a graph?

Statistics, Data Analysis, and Graphing

15 Graphs are a perfect representation of the data. Do you agree or disagree with this statement?

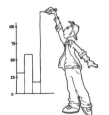

Statistics, Data Analysis, and Graphing

16 What kinds of data could you collect about your daily activities? What kinds of things do you think you would observe about yourself by analyzing the data?

Higher-Level Thinking Questions for Secondary Mathematics
Kagan Publishing • 1 (800) 933-2667 • www.KaganOnline.com

Statistics, Data Analysis, and Graphing

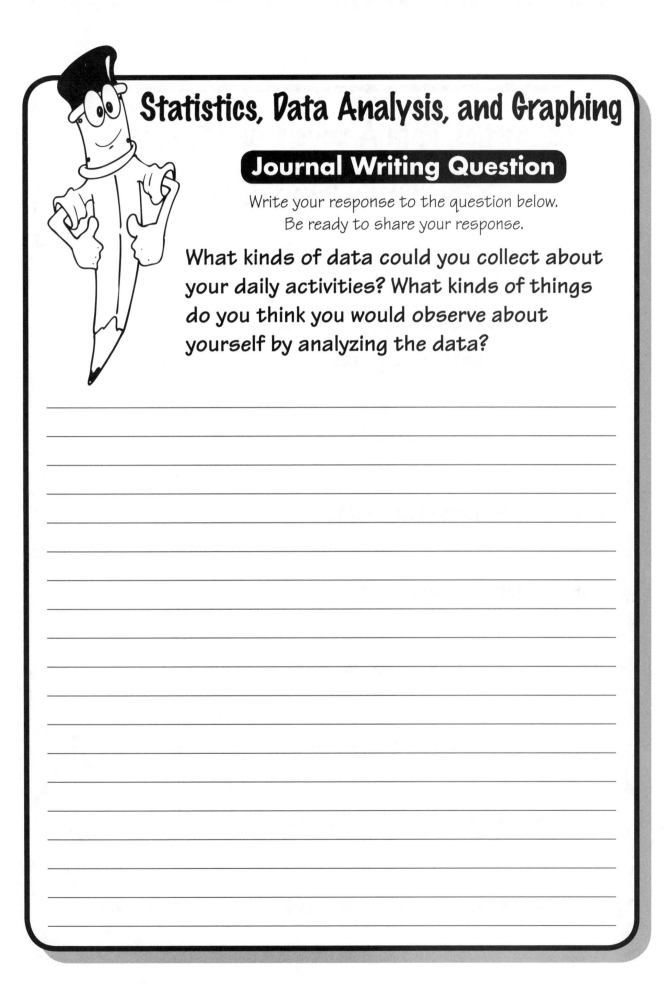

Journal Writing Question

Write your response to the question below.
Be ready to share your response.

What kinds of data could you collect about your daily activities? What kinds of things do you think you would observe about yourself by analyzing the data?

Statistics, Data Analysis, and Graphing

Question Starters

Use the question starters below to create complete questions.
Send your questions to a partner or to another team to answer.

1. Why are graphs
_____

2. What kinds of patterns can data
_____

3. How can graphs help predict
_____

4. When are graphs the best way to
_____

5. How can data analysis be useful in
_____

6. How are statistics and data analysis
_____

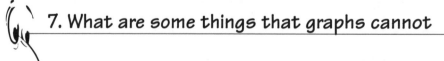

7. What are some things that graphs cannot
_____

8. To create a graph, what do you
_____

Higher-Level Thinking Questions for Secondary Mathematics
Kagan Publishing • 1 (800) 933-2667 • www.KaganOnline.com

Algebra: Integers and Integer Operations

higher-level thinking questions

"I had a feeling once about Mathematics—that I saw it all. Depth beyond depth was revealed to me—the Byss and Abyss. I saw—as one might see the transit of Venus or even the Lord Mayor's Show—a quantity passing through infinity and changing its sign from plus to minus. I saw exactly why it happened and why the tergiversation was inevitable but it was after dinner and I let it go."

— Sir Winston Spencer Churchill

Higher-Level Thinking Questions for Secondary Mathematics
Kagan Publishing • 1 (800) 933-2667 • www.KaganOnline.com

Algebra: Integers and Integer Operations

1 Integers can be positive or negative. What are some examples of negative values that are advantageous to have in real life?

Algebra: Integers and Integer Operations

2 How does the statement "One person's loss is another person's gain" apply to integers? Give specific examples.

Algebra: Integers and Integer Operations

3 What is the opposite of adding two negative integers?

Algebra: Integers and Integer Operations

4 What generalizations can you make about working with 0 and 1 as integers?

Algebra: Integers and Integer Operations

5 What are some patterns that help you predict the products and quotients of integer pairs?

Algebra: Integers and Integer Operations

6 How is absolute value related to integers?

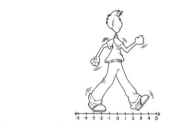

Algebra: Integers and Integer Operations

7 What are some questions you ask yourself when finding the difference of two integers?

Algebra: Integers and Integer Operations

8 The set of all integers is usually denoted by **Z**, Zahlen, the German word for "number." How could you remember the meaning of **Z**?

Higher-Level Thinking Questions for Secondary Mathematics
Kagan Publishing • 1 (800) 933-2667 • www.KaganOnline.com

Algebra: Integers and Integer Operations

9 What are some ways that integers can be compared and ordered?

Algebra: Integers and Integer Operations

10 Meteorologists use integers to indicate a positive or negative change in temperature. What are some additional examples of integers being used to show positive or negative change?

Algebra: Integers and Integer Operations

11 Integers consist of the positive natural numbers (1, 2, 3, …), the negative natural numbers (-1, -2, -3, …), and the number zero (0). What are some values that are not integers?

Algebra: Integers and Integer Operations

12 Infinite means "having no limits." Why do you think integers are considered "countably infinite"?

Algebra: Integers and Integer Operations
Question Cards

13 Negative numbers are useful to describe values on a scale that go below zero such as temperature. Besides temperature, what else could negative numbers represent?

14 A plus sign and a minus sign can be used to represent positive and negative integers. What are some other ways you can symbolize the difference between positive and negative integers?

15 You are inventing logical, easy-to-recall, one-word names for positive integers and negative integers. What are the names? Why do they make sense?

16 Positive and negative numbers behave differently in mathematical operations. How do 5 and -3 behave differently when added than when multiplied?

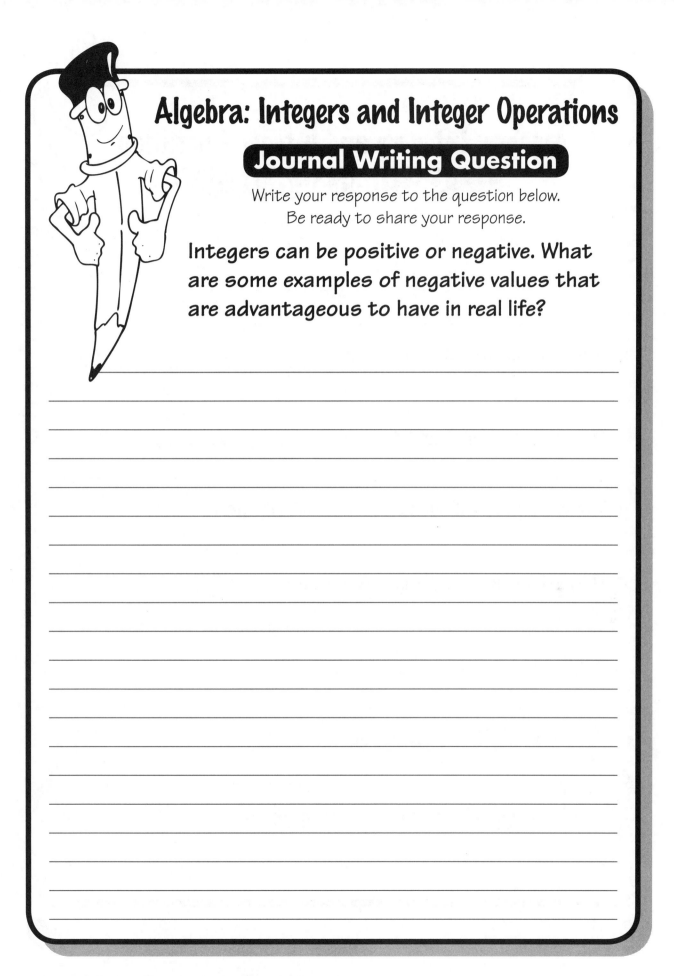

Algebra: Integers and Integer Operations

Journal Writing Question

Write your response to the question below.
Be ready to share your response.

Integers can be positive or negative. What are some examples of negative values that are advantageous to have in real life?

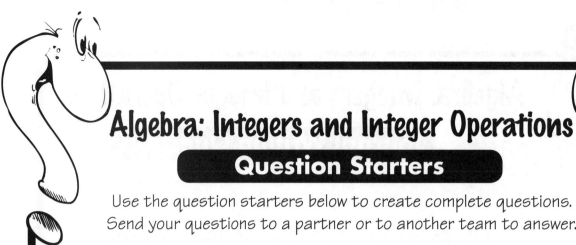

Algebra: Integers and Integer Operations
Question Starters

Use the question starters below to create complete questions.
Send your questions to a partner or to another team to answer.

1. How are integers like

2. When is the sum of two integers

3. Why aren't integers used

4. Why is the sign of the product of two integers

5. How are positive and negative integers

6. What integer

7. Where are integers found in

8. In which professions are integers

Higher-Level Thinking Questions for Secondary Mathematics
Kagan Publishing • 1 (800) 933-2667 • www.KaganOnline.com

Algebra: Solving Equations and Inequalities

higher-level thinking questions

"Inequality is the cause of all local movements.

— Leonardo da Vinci"

Higher-Level Thinking Questions for Secondary Mathematics
Kagan Publishing • 1 (800) 933-2667 • www.KaganOnline.com

Algebra: Solving Equations and Inequalities
Question Cards

Algebra: Solving Equations and Inequalities

1 Equations and inequalities have variables, or letters, to represent unspecified values. What are some variables in your life? What causes the variables to change over time?

Algebra: Solving Equations and Inequalities

2 What are some visual representations for equations and inequalities?

Algebra: Solving Equations and Inequalities

3 Learning how to solve math problems gives you insight to solve other problems you encounter in life. Do you agree or disagree? Why?

Algebra: Solving Equations and Inequalities

4 When solving equations, the input affects the output. When are some times that you have input into something that ultimately affected the output?

Algebra: Solving Equations and Inequalities
Question Cards

Algebra: Solving Equations and Inequalities

5 How is the order of operations related to the steps in solving two- and multi step equations? How is it related to checking those equations?

Algebra: Solving Equations and Inequalities

6 What are some real-life situations in which the resulting graph of the equation would be a line with a negative slope?

Algebra: Solving Equations and Inequalities

7 The dictionary definition of expression is a "word, phrase, look, or intonation that conveys meaning." In what ways is the mathematical meaning for expression an interpretation of the dictionary's definition?

Algebra: Solving Equations and Inequalities

8 You have $5 to spend on lunch at a fast food restaurant. How is selecting your meal and drink solving an inequality problem? What consideration influences your decision?

Higher-Level Thinking Questions for Secondary Mathematics
Kagan Publishing • 1 (800) 933-2667 • www.KaganOnline.com

Algebra: Solving Equations and Inequalities

9 What are some ways that bowlers and golfers of differing skill levels can use algebra and equations to equalize their scores?

Algebra: Solving Equations and Inequalities

10 Some mobile phone plans sell a fixed number of minutes for a set price, and additional minutes at a rate per minute. What are some other real-life situations in which a fixed rate is combined with an undetermined variable rate?

Algebra: Solving Equations and Inequalities

11 What are some characteristics of an equation that affect the appearance of the resulting graph?

Algebra: Solving Equations and Inequalities

12 What are some ways you can classify the graph of the equation of a line?

Algebra: Solving Equations and Inequalities

13 How is the process of solving an equation like solving an inequality? How is it different?

Algebra: Solving Equations and Inequalities

14 How is solving an equation like solving a puzzle? How is it different?

Algebra: Solving Equations and Inequalities

15 To solve equations, you perform operations to isolate a variable. Where else do people try to isolate variables to solve the unknown?

Algebra: Solving Equations and Inequalities

16 What connections can you make between the equation of a line and a line graph?

Higher-Level Thinking Questions for Secondary Mathematics
Kagan Publishing • 1 (800) 933-2667 • www.KaganOnline.com

Algebra: Solving Equations and Inequalities

Journal Writing Question

Write your response to the question below.
Be ready to share your response.

Learning how to solve math problems gives you insight to solve other problems you encounter in life. Do you agree or disagree? Why?

Algebra: Solving Equations and Inequalities

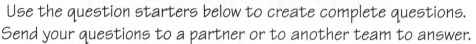

Question Starters

Use the question starters below to create complete questions.
Send your questions to a partner or to another team to answer.

1. How are equations and inequalities

_____

_____

2. What if an equation

_____

_____

3. Why are equations

_____

_____

4. When evaluating an equation,

_____

_____

5. When are inequalities

_____

_____

6. How are inequalities related to

_____

_____

7. If inequalities were

_____

_____

8. What reasons might explain why inequalities

_____

_____

Higher-Level Thinking Questions for Secondary Mathematics
Kagan Publishing • 1 (800) 933-2667 • www.KaganOnline.com

Geometry
Concepts

higher-level thinking questions

"Guided only by their feeling for symmetry, simplicity, and generality, and an indefinable sense of the fitness of things, creative mathematicians now, as in the past, are inspired by the art of mathematics rather than by any prospect of ultimate usefulness.

— Eric Temple Bell

Geometry Concepts
Question Cards

1 The first video game programmers used quadrilaterals, such as rectangles, squares, and trapezoids, in designing video games. What are some other geometric concepts that might be applied when designing video or computer games?

2 The hands of an analog (clock face) clock form an acute angle at 3:05, a straight angle at 6:00, and an obtuse angle at 3:40. What are some other examples of right, acute, straight, and obtuse angles on the clock face?

3 The term geometry comes from two Greek words meaning "earthly measurement." Why do you think this type of mathematics was named as such, and how has geometry changed today?

4 Your friend has just become an architect. You want to buy her a tool essential to her success. What tools would you purchase for her and how would she use them? Explain.

Geometry Concepts
Question Cards

Geometry Concepts

5 What are some of the relationships between angle pairs? What are some real-life instances in which these angle relationships are important to know?

Geometry Concepts

6 Is geometry more valuable to an architect, engineer, or astronomer? Why?

Geometry Concepts

7 Right angles can be found in picture frames, chalkboards, and carpets. What are some places where acute or obtuse angles can be found? Where can obtuse angles be found?

Geometry Concepts

8 You are planning to construct a pair of perpendicular or a pair of parallel lines. What tools would you need? What steps would you take to complete the process?

Higher-Level Thinking Questions for Secondary Mathematics
Kagan Publishing • 1 (800) 933-2667 • www.KaganOnline.com

Geometry Concepts
Question Cards

Geometry Concepts

9 Triangles can be classified using angles or side measures. A triangle may be described as acute isosceles or obtuse scalene. You are renaming triangles with one-word descriptions. What are the new names? How did you arrive at your one-word descriptors?

Geometry Concepts

10 Many flags and pennants have angular designs. What are some other examples of everyday items that have angular designs?

Geometry Concepts

11 You have been asked to rename one of the quadrilaterals (square, parallelogram, rectangle, rhombus, trapezoid). Which quadrilateral would you rename, and what would be its new name? Justify your decision.

Geometry Concepts

12 You are creating a street sign that warns drivers to stop for school buses while loading or unloading. What geometric shape would you use? Explain your decision.

Geometry Concepts
Question Cards

13 Choose a geometric term or concept, such as tessellation, diagonal, or polygon. How might you teach the concept or term to a seven-year-old?

14 What are some geometric patterns that can be found in nature? Describe the patterns and where you see them.

15 You are fascinated with the geometric vision exemplified by bridges such as the Golden Gate Bridge in San Francisco. What are some questions related to geometry or measurement that you would ask a bridge designer?

16 Are you more like a circle, a square, a triangle, or a pentagon? Why?

Geometry Concepts

Journal Writing Question

Write your response to the question below.
Be ready to share your response.

The term geometry comes from two Greek words meaning "earthly measurement." Why do you think this type of mathematics was named as such, and how has geometry changed today?

Geometry Concepts
Question Starters

Use the question starters below to create complete questions.
Send your questions to a partner or to another team to answer.

1. When are lines and angles

2. In nature, geometry is found

3. Why are geometric transformations

4. When, other than in geometry, are prefixes

5. How are geometric patterns

6. For what careers and professions is geometry

7. Why are geometric figures

8. What geometric terms

Higher-Level Thinking Questions for Secondary Mathematics
Kagan Publishing • 1 (800) 933-2667 • www.KaganOnline.com

Geometry and Measurement

higher-level thinking questions

The so-called Pythagoreans, who were the first to take up mathematics, not only advanced this subject, but saturated with it, they fancied that the principles of mathematics were the principles of all things.

— Aristotle

Higher-Level Thinking Questions for Secondary Mathematics
Kagan Publishing • 1 (800) 933-2667 • www.KaganOnline.com

Geometry and Measurement
Question Cards

1 The Pythagorean Theorem ($a^2 + b^2 = c^2$) states that the squares of the two shorter sides, or legs, of a right triangle are equal to the square of the triangle's longest side, or hypotenuse. What are some geometry and algebra concepts that you must know in order to understand the Pythagorean Theorem?

2 What are some applications in real life or in nature of the Pythagorean Theorem?

3 A homeowner wants to fence in a garden area on his property. What are some ways that he could maximize the area of the garden and minimize the amount of fencing required?

4 What is the relationship among the area of a triangle, the area of a rectangle, and the area of a parallelogram?

Geometry and Measurement
Question Cards

Geometry and Measurement

5 The areas of states can be estimated by combining the areas of geometric shapes. Choose a state. How would you break apart the shape of the state to estimate its area?

Geometry and Measurement

6 What is more important for everyday life: How to calculate area, volume, or surface area? Defend your choice.

Geometry and Measurement

7 What is the value of π (pi) and what does it tell us about the relationship between the circumference and diameter of a circle?

Geometry and Measurement

8 How are surface area and volume related? How would you compare the processes for finding these measurements?

Higher-Level Thinking Questions for Secondary Mathematics
Kagan Publishing • 1 (800) 933-2667 • www.KaganOnline.com

Geometry and Measurement
Question Cards

Geometry and Measurement

9 "If the volumes of two cylinders are equal, then their surface areas must be equal." Do you agree or disagree with this statement? What experiment would you devise to prove that you are correct?

Geometry and Measurement

10 You have been nominated to rename geometric measurements to make them simpler and easier to remember. What names will you choose to replace volume, perimeter, area, and surface area?

Geometry and Measurement

11 You buy 5 posters that you want to place on your bedroom wall. In positioning the posters so they are visually pleasing, would you use precise measurements or estimates? Explain.

Geometry and Measurement

12 You have devised a new calculator key that automatically calculates the surface area of a figure. What information would you have to enter into the calculator? What would the symbol look like for your new key?

Geometry and Measurement
Question Cards

Geometry and Measurement

13 In a game of charades, you must demonstrate finding the surface area of a rectangular prism. What are some things you would do to convey this idea?

Geometry and Measurement

14 You are creating a new food product. What are some questions you must ask yourself before deciding whether to package the product in a rectangular prism or a cylinder?

Geometry and Measurement

15 What are some professions that utilize skills and concepts in measuring 2- and 3-dimensional figures? Which specific skills and concepts are required most often?

Geometry and Measurement

16 There are specific formulas for the volumes of solid figures such as rectangular prisms, cylinders, cones, and pyramids. What are some generalizations you can make about finding the volumes of these figures?

Higher-Level Thinking Questions for Secondary Mathematics
Kagan Publishing • 1 (800) 933-2667 • www.KaganOnline.com

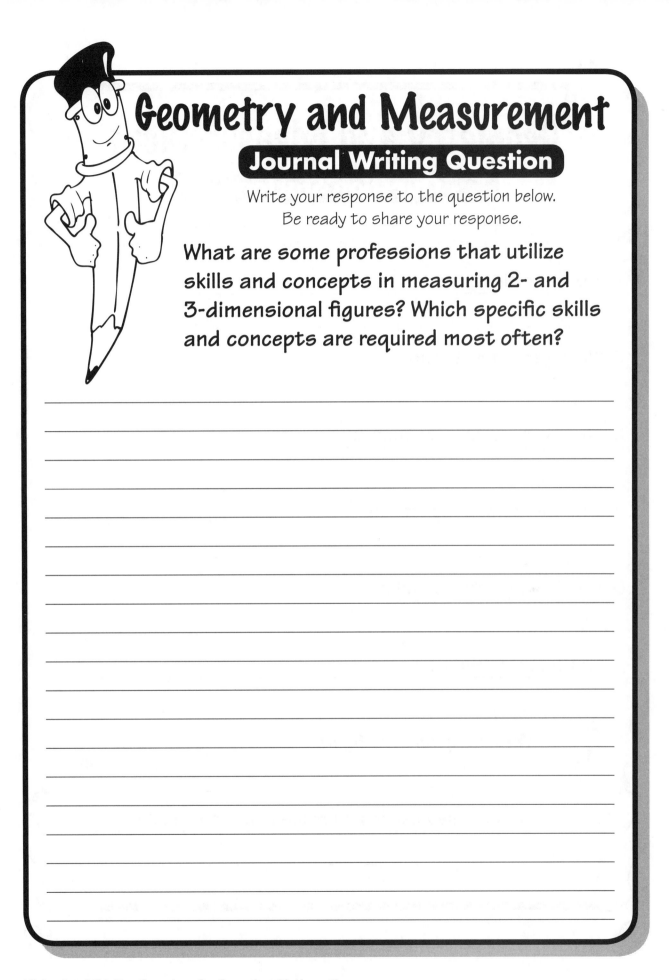

Geometry and Measurement

Journal Writing Question

Write your response to the question below.
Be ready to share your response.

What are some professions that utilize skills and concepts in measuring 2- and 3-dimensional figures? Which specific skills and concepts are required most often?

Geometry and Measurement
Question Starters

Use the question starters below to create complete questions.
Send your questions to a partner or to another team to answer.

1. How are solid figures

2. What geometric terms

3. When is a precise measurement

4. Why are geometric patterns

5. How is a figure's area related to

6. Which geometric solid

7. When are geometric figures

8. How do geometric and measurement formulas

Higher-Level Thinking Questions for Secondary Mathematics
Kagan Publishing • 1 (800) 933-2667 • www.KaganOnline.com

Probability

higher-level thinking questions

"To us probability is the very guide of life.

— Bishop Butler

Higher-Level Thinking Questions for Mathematics
Kagan Publishing • 1 (800) 933-2667 • www.KaganOnline.com

Probability
Question Cards

1 What are some ways that probability concepts are applied when calculating health or life insurance costs?

2 Some people who believe that "history repeats itself" use events of the past to predict the probability of events in the future. What are some examples of history repeating itself in your life's experiences?

3 You are the manager of a successful restaurant. What are some ways that probability concepts help you run a better business?

4 The probability of an occurrence ranges from zero (impossible) to one (certain). What real-life occurrences have a probability of exactly one-half? Explain your reasoning.

Probability
Question Cards

Probability

5 You are playing a card or dice game with an opponent who has not yet studied probability. Do you have the winning advantage? Explain.

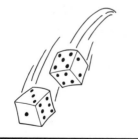

Probability

6 How would you complete the analogy, "Probability is to a combination lock as ____ is to ____"? Explain your thinking.

Probability

7 What are some examples of real-life experiences for which probability is not a factor?

Probability

8 What role does probability play in gambling?

Probability
Question Cards

Probability

9 What are some real-life instances in which a lower probability is associated with a more positive outcome? Give specific examples.

Probability

10 What are some ways you could express the probability of an event using a drawing or symbol?

Probability

11 What are some observations that may help you predict the probability of an event?

Probability

12 A probability experiment involves using a spinner. What are some ways to design the experiment so that it is fair?

Probability

13 How are permutations and combinations alike? How are they different?

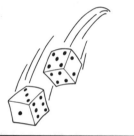

Probability

14 What is the significance of probability when taking a selected response (multiple choice) test?

Probability

15 "In a presidential election, the probability of the outcome can affect the actual outcome." Do you agree or disagree with this statement? Explain your reasoning.

Probability

16 What qualities or characteristics are common to games that involve probability? How are they different from games that do not involve probability?

Higher-Level Thinking Questions for Mathematics
Kagan Publishing • 1 (800) 933-2667 • www.KaganOnline.com

Probability

Journal Writing Question

Write your response to the question below.
Be ready to share your response.

Some people who believe that "history repeats itself" use events of the past to predict the probability of events in the future. What are some examples of history repeating itself in your life's experiences?

_____
_____
_____
_____
_____
_____
_____
_____
_____
_____
_____
_____
_____
_____
_____
_____
_____
_____
_____

Probability

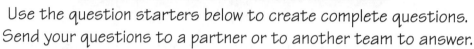

Use the question starters below to create complete questions.
Send your questions to a partner or to another team to answer.

1. How is probability similar to

_____

2. What skills and concepts does probability

_____

3. What factors affect the probability of

_____

4. What causes the probability of an event

_____

5. In what ways are games and probability

_____

6. When finding the probability of an event, why is

_____

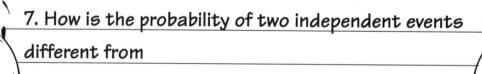

7. How is the probability of two independent events

different from

_____

8. How can probability be used to

_____

Higher-Level Thinking Questions for Mathematics
Kagan Publishing • 1 (800) 933-2667 • www.KaganOnline.com